Horticultural machinery

(Third edition)

Frontispiece. A specialised turf-care tractor.
By permission of Sisis Equipment (Macclesfield) Ltd.

Horticultural machinery

M. F. J. Hawker and J. F. Keenlyside

 Longman
Scientific &
Technical

Longman Scientific & Technical
Longman Group UK Limited
Longman House, Burnt Mill, Harlow
Essex CM20 2JE, England
and Associated Companies throughout the world.

First published 1971 by Macdonald & Co. (Publishers) Ltd.
Second edition 1977 by Longman Group Ltd.
Third edition 1985
Reprinted 1989

British Library of Cataloguing in Publication Data
Hawker, M. F. J.
    Horticultural machinery.-3rd ed.
    1. Horticultural machinery
    I. Title   II. Keenlyside, J. F.
    635'.028   S678.7
    ISBN 0-582-40807-5

Library of Congress Cataloging in Publication Data
Hawker, M. F. J. (Michael Frederick James), 1932–
    Horticultural machinery.

    Includes index.
    1. Horticultural machinery.   I. Keenlyside, J. F.
(John Frederick), 1942– .II. Title.
S678.7.H38   1985   635'.028   84-20170
ISBN 0-582-40807-5

Set in Linotron 202 9/10pt Memphis Light
Produced by Longman Group (FE) Ltd
Printed in Hong Kong

Contents

Preface to the third edition

Considerable developments in horticultural mechanization have taken place in the fourteen years since the first edition of *Horticultural Machinery* was prepared. To take account of this, major revisions of several chapters have been made, and new sections on brakes, governors, hydrostatic transmission, power steering, and rotary sieves, among others, have been added. More drawings have been included as these are considered to be one of the best ways of explaining simply to the reader how a mechanism works, or how it is constructed. They are drawn as simplified diagrams and are not intended to show every detail.

The contents of the book complement the current Phase I and Phase II syllabuses in Horticulture, and the new sections will be of particular assistance to those studying the Phase II options in Greenkeeping and Sports Turf Management. The book will also be of value to those following National Certificate and National Diploma courses in Amenity and Commercial Horticulture and will serve as an introduction to higher-level courses.

As in the previous edition, SI has been adopted as the standard system of units, but as imperial units are still being used in many operator instruction books and on equipment such as tyre pressure gauges, imperial equivalents are included in parentheses after SI values for important settings and adjustments.

List of figures

1 Materials and workshop tools

In the production of modern horticultural machinery a wide range of materials is used. These can be divided into *metals* and *non-metals*. The metals used can further be divided into *ferrous metals* and *non-ferrous metals*.

Ferrous metals

Ferrous metals all contain iron. The different metals in this group contain carbon and other elements which determine their properties. The common ones are described below.

Cast iron (contains 3 to 3.5% carbon)

Cast iron is used in the production of castings such as cylinder blocks and gearboxes. As it becomes very fluid when hot, it is particularly suitable for complicated castings. The rate at which the cast iron cools affects the property of the casting. Cast iron will withstand compressive forces but easily fractures if dropped. To improve its wearing properties small quantities of nickel or chromium can be added.

Chilled cast iron

When cast iron is cooled rapidly a hard layer is formed on the surface of the casting. Chilling also enables a casting to be machined easily.

Malleable iron (contains 0.5 to 1.0% carbon)

Malleable iron is capable of withstanding shock loads and so can be shaped with a hammer. It is not brittle like cast iron and it is used for making harrow tines and mower fingers which can be straightened with a hammer without fear of breakage.

Wrought iron (contains 0.02 to 0.03% carbon)

Wrought iron is produced by removing most of the carbon from cast iron. It is easily worked when hot and is widely used for ornamental ironwork such as gates, railings and plant stands.

Steel

Steel is widely used in the construction of horticultural machinery. There are different types of steel which depend on the quantity of carbon and the other alloying elements in them.

Mild steel

Mild steel has a very low carbon content (0.1 to 0.25%). It is used as angle iron and plate on horticultural machines. It can be welded easily, but is soft and therefore is not very resistant to wear and corrodes very easily.

Medium-carbon steel

Medium-carbon steel contains more carbon than mild steel (0.25 to 0.5%). It has good resistance to wear and so is used for plough beams, soil-working parts and mower blades.

High-carbon steel

High-carbon steel contains 0.5 to 1.5 per cent carbon and is very hard. It is used in the production of tools such as twist drills, punches, and hacksaw blades.

Other metals, such as nickel, tungsten, and manganese, are often alloyed with steel to give it special qualities for specific purposes. For instance high-speed cutting tools and spanners are made out of alloys containing these metals.

All ferrous metals are subject to corrosion and must be protected. Mild steel will rust overnight if not protected. Red oxide or bitumastic paints help to prevent rusting in the atmosphere, but where steel is in contact with water galvanizing provides better protection.

Non-ferrous metals

Non-ferrous metals do not include iron. Many of them are resistant to corrosion and are used where this is a problem. They tend to be more expensive than iron and steel so are not used in such great quantities.

Copper

Copper is an excellent conductor of heat and electricity and is used in electrical appliances. It is also used extensively for hot-water supplies as copper tube can be bent and soldered readily.

Lead

Lead is a heavy metal and is very soft. It is used in batteries, for flashings on roofs and in solder. Lead is no longer used for water pipes due to its cost, but is still used as a basis for some paints.

Tin and zinc

Tin and zinc are only used as protective coatings on steel and are not used as pure metals in horticulture. Cans dipped in molten tin are used in the food industry. Where steel fittings are likely to rust, they can be galvanized by dipping them into molten zinc.

Brass

Brass is an alloy of copper and zinc and may also contain some lead. Brass is used in bearings and also for many electrical connections. Because it is resistant to chemical corrosion it is sometimes used in sprayer pumps and nozzles.

Aluminium

Aluminium is a very light metal, and alloys containing a high proportion of this metal are widely used in the construction of glasshouses. It is resistant to corrosion and its use has considerably reduced the maintenance of glasshouses. Aluminium is also used for the production of lightweight fittings, and in the manufacture of engine cylinder blocks because of its good casting properties.

Plastics

Plastics are widely used for the production of plant pots and seed trays and as these are easily cleaned have made glasshouse hygiene very much simpler. Plastics have also been developed to resist ultraviolet light which tended to make pots and trays brittle when used in a glasshouse. Double glazing of glasshouses is simple with polythene sheeting. Black polythene is used for mulching and for controlling light for all-year-round chrysanthemum production. Plastic piping is widely used for irrigation, drainage and water supplies because one of its main advantages is that it is completely corrosion resistant. Plastic components are used in spraying equipment because they are resistant to attack by spray chemicals. Watering cans and buckets are commonly made of plastics and special plastic surfaces are now being produced for all-weather sports pitches.

Nylon

Nylon is harder than plastics and can withstand higher temperatures without melting. It is used to make caps to protect chrysanthemum blooms and spun nylon is used to make sprout bags. Nylon surfaces working in contact do not require lubrication and so it is used for small gearwheels, chain tensioners and bushes on many machines.

Workshop tools

In order to carry out simple maintenance tasks on horticultural machinery, it is necessary to be able to select the correct tools and to use them properly.

Spanners and sockets

Spanners and sockets are used to tighten and undo nuts and bolts. Unfortunately, nut sizes and bolt threads are not standardized and four common ranges are in use; spanners will seldom fit nuts and bolts from other ranges and nuts from one range will seldom fit bolts of another. The ranges are:

1. Whitworth.
2. British Standard Fine.

One set of spanners will fit both these ranges but the nuts are not interchangeable. The size stamped on the jaw of the spanner is the bolt diameter. In recent years they have been largely superseded on new equipment by:

3. Unified (A/F)

in which the size is measured across the flats in inches; and

4. Metric

in which the size is measured across the flats in millimetres.

Metric sizes and threads in keeping with SI units are rapidly becoming the standard in this country. Ultimately this will make the correct selection of a spanner much easier and in time other size ranges will disappear.

The correct size of spanner to fit the nut must always be used, otherwise the corners of the nut will be damaged making it difficult to move it even with the correct size of spanner. A length of pipe should not be slipped over a spanner to lengthen it and increase the leverage that can be applied to a nut, as this will strain the spanner jaws and may break the bolt or strip its thread.

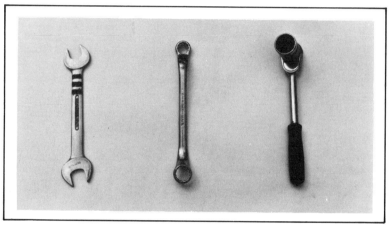

Fig. 1.1 A selection of spanners.
From left to right – an open-ended spanner, a ring spanner, and a socket with a ratchet handle.

The three main types (see Fig. 1.1) used are:

1. open-ended spanners;
2. ring spanners;
3. sockets.

Open-ended spanners and ring spanners have jaws or rings at each end which normally fit different sizes of nut.

Open-ended spanners

These are used where it is impossible to fit a.ring spanner because all sides of the nut are not accessible. The jaws at each end are angled to the central grip so that the spanner can be used in two different positions when moving a nut in a restricted space.

Ring spanners

These fit right round all six sides of the nut and will not slip off or

4

damage it. It is much safer to use this type of spanner when moving a stubborn nut.

Sockets

Fitted right over the nut these can be used with a selection of handles such as a ratchet or a brace. Different accessories enable nuts in very awkward places to be moved. Although expensive, a good set of sockets with a range of handles is essential in any workshop. Extra-long sockets are available which fit right over a sparking plug so that it can be removed without damage. Some of these plug sockets have rubber inserts in the 'top' end to protect the ceramic insulation.

Wrenches

Various types of adjustable wrenches are available for the horticultural workshop. Stillsons, which are designed for pipe work, and water-pump pliers must be used with care otherwise damage to the nut edges will occur.

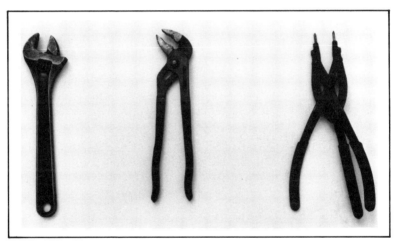

Fig. 1.2 Workshop tools.
From left to right, an adjustable wrench, water-pump pliers, and circlip pliers.

Chisels

Chisels are used for cutting metal (see Fig. 1.3). They are often referred to as cold chisels as they cut the metal when cold. They are made out of medium-carbon steel which has been hardened and tempered. Chisels have a hexagonal shank which is flattened and ground to a cutting edge. The angle of the chisel edge depends on what metal it is to cut, but 60 degrees is common. When sharpening a chisel on the grindstone, care must be taken not to overheat it or it will lose its temper and will become soft. After a while the other end of a chisel will develop a 'mushroom' head due to continual hammering. This must be removed

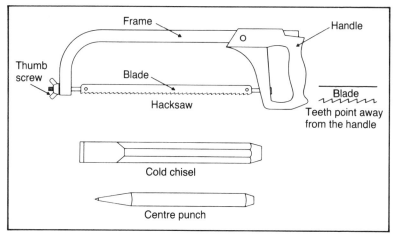

Fig. 1.3 Workshop tools.

with a grindstone otherwise pieces of the chisel head may fly off and hit the operator.

Punches

The centre punch (Fig. 1.3) has a shank which comes to a point and is used with a hammer to make a mark to start a twist drill when drilling metal. Other punches have either tapered or parallel shanks with a flat end and are used for removing rivets and split pins.

Hacksaws

The hand hacksaw (Fig. 1.3) has a place in every workshop. It consists of a frame which can be adjusted to take either 250 or 300 mm long blades.

Hacksaw blades can be obtained with different numbers of teeth per millimetre. Common grades have 18, 24, or 32 teeth per 25 mm. When cutting soft metals a coarse-toothed blade must be used so that the teeth do not become clogged. Fine-toothed blades are used when the cutting surface is thin so that at least three teeth are in contact with the work all the time to prevent the blade from jamming. When fitting a blade into a hacksaw, the teeth should point away from the handle so that cutting takes place on the forward stroke. The blade is tensioned by means of a thumbscrew, and is correctly adjusted when a high-pitched note is produced when the blade is plucked.

The hacksaw must be held correctly to obtain the best results. The right-handed operator holds the handle in his right hand and steadies the front of the frame with his left. The work to be cut must be held securely in a vice and the first cut made with gentle forward pressure. A steady rate of 50 strokes per minute will make the cutting easy. If a blade breaks in the middle of a cut, then a new cut must be started from the other side, because the new blade, which will be wider, will not pass down the original slot.

Files

A file is used to remove small quantities of metal. It is a cutting tool and contains rows of teeth which point from the handle and, like the hacksaw, cut on the forward stoke. One end of the file is drawn out into a spike called the tang, and a file must always be used with a wooden or plastic handle on the tang.

Files are grouped according to their cross-sections – flat, round, half-round, triangular. and square can be obtained. Files are classified according to the number of teeth per 25 mm, for instance, a bastard file has 30 teeth, a second cut 40 teeth and a smooth cut 50 teeth. Like the hacksaw blade, a coarse file is used for soft metal and a fine file for harder metal.

When filing it is important that the file is cleaned regularly with a wire brush to remove all traces of metal from the teeth. Chalk rubbed on the file also helps to prevent the teeth from clogging. The metal to be filed must be held securely and the file held correctly by the operator. Pressure is applied during the forward cutting stroke and the file is lifted or eased on the return stroke. The operator must hold the work to be filed at a comfortable height and stand so that his weight is taken equally on both feet. The file must not be rocked when filing but held level throughout the stroke. The filing rate is about 50 strokes per minute.

Files are expendable and must be discarded when worn out. They should not be thrown together in a box after use as this reduces their life. Instead they should be stored in a rack, or wrapped in paper.

Hammers

Ball-pein hammers are most commonly used in engineering and are classified according to their weight. They are used in conjunction with chisels .and punches as described earlier. The steel hammer head is fastened on to a wooden shaft and the head has one flat face, the other being rounded into a ball which is used for riveting. Hammer shafts are usually made of hickory or ash. When gentle hammer blows are required in fitting, a combination hammer is recommended. These have soft faces, one of copper the other of hide, which will not damage the surfaces being hammered and can be replaced when worn. Nylon-headed hammers are also used for this purpose.

Pliers

Pliers are used for holding when a good grip is essential. Engineers pliers have serrated jaws with flat ends. Also included in the jaws are a pair of cutting edges which are useful for cutting wire or thin metal. Tapered-nose pliers have much longer jaws which taper towards the end and are useful for fine work. Other special pliers such as circlip pliers (Fig. 1.2) for removing and fitting external and internal circlips are useful additions to the workshop.

Screwdrivers

Most engineers screwdrivers have plastic handles which are insulated and tested to withstand 10 000 V (volts). Other screwdrivers have wooden handles into which the steel blade is fitted. The blade of this type is square in section so that a spanner can be used to provide more leverage when unscrewing a tight screw. Screwdrivers are classed according to their blade length, 150, 200 and 250 mm being common lengths.

Phillips screwdrivers have a special blade shaped to fit screws with a slot shaped like a cross, and 'Pozidrive' screwdrivers with a similar blade but with a more straight-sided slot are also available. This type of slot is often found on self-tapping screws. When loosening a screw always use a screwdriver with a blade that reaches to the bottom of the slot. This is particularly necessary with a Phillips or Pozidrive screw to avoid damage to the edge of the slot, making removal difficult. The screwdriver should be kept in line with the screw and sufficient force applied to keep the end of the blade in the slot.

Feeler gauges

When very small gaps have to be adjusted, such as contact-breaker gaps and sparking plug gaps, feeler gauges are necessary (Fig. 1.4). They consist of a set of specially rolled steel blades of different thicknesses, measured in hundredths of a millimetre or thousandths of an inch. Different blades can be used in combinations to obtain a desired thickness. Feeler gauges must be treated with care and the blades slightly oiled occasionally to prevent them becoming rusty.

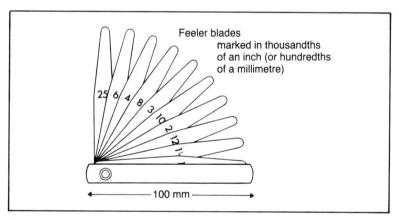

Feeler blades marked in thousandths of an inch (or hundredths of a millimetre)

100 mm

Fig. 1.4   Set of feeler gauges.

Nuts and bolts

Nuts and bolts are a very common way of fastening components together. They also have the advantage of being easy to remove. Threads can be classified into coarse threads and fine threads.

1. *Coarse threads*: British Standard Whitworth (BSW); Unified Coarse (UNC); British Standard Pipe (BSP); and Metric Coarse. These are used in castings or where thread strength is important.

2. *Fine threads*: British Standard Fine (BSF); Unified Fine (UNF); Metric and British Association (BA). These are used where fine adjustments are required or vibration of components is a problem.

Neither coarse-thread nuts nor fine-thread nuts are necessarily interchangeable.

Questions

1. Name three ferrous metals.
2. Name three non-ferrous metals.
3. Name three ways in which metals can be prevented from rusting.
4. What is an alloy?
5. What advantages have plastics over steel?
6. What system is to be used in this country to standardize the nut sizes and bolt threads?
7. What advantages have ring spanners over open-ended spanners?
8. Why must coarse-toothed hacksaw blades be used when cutting soft metals?
9. What is a suitable filing rate?
10. What is a combination hammer and when is it used?
11. Name three coarse threads.
12. Name three fine threads.

2 Mechanics

The design of all mechanical devices used in horticulture is based upon a series of fundamental mechanical laws and principles. A knowledge of these simple principles makes it much easier to appreciate and understand how horticultural machines work. Thus a number of basic definitions and principles have been outlined in this chapter. In addition, a number of commonly used terms have been defined.

SI units

SI units are now being used in many branches of industry, including horticulture, and are encountered increasingly in everyday life. The continuing change-over from imperial units to SI is still recognized policy, and so SI units are used almost exclusively throughout this book. This internationally-accepted system of units is based upon a small number of fundamental units from which all other units can be derived. The more important basic units, and a number of useful units derived from them which are particularly relevant in horticulture, are given below with their standard abbreviations. Note how the squared and cubed values are indicated, and that no full stops are used following abbreviations.

Measurement	Unit	Symbol
Length	metre	m
Area	square metre	m^2
	hectare	ha
Volume	cubic metre	m^3
	litre	ℓ
	(1 ha = 10 000 m^2)	

To indicate very large or very small quantities and dimensions a series of prefixes can be added to any unit. The commonly used ones are as follows:

Prefix	Symbol	Value
milli-	m	$X \dfrac{1}{1\ 000}$
kilo-	k	X 1 000
mega-	M	X 1 000 000

for example, 1 kilometre (km) = 1 000 metres (m) = 1 000 000 millimetres (mm). Other SI units occur elsewhere in the text and will be explained when they are introduced.

A force

The term **force** is widely used. The definition of a force is that it changes, or tries to change, a body's state of rest or uniform motion in a straight line. In simple terms this means that whenever a stationary object is set into motion, or whenever a moving object is accelerated or decelerated, a force is responsible for the change.

Thus a tractor exerts a force on an implement when it pulls it along. Similarly, the driver exerts a force when he pushes the gear lever into gear.

If the driver tries to pull the implement himself he may not have much success but, even if he cannot move it, he is still exerting a force on the implement which tends to move it.

The unit of force is the newton (N), but as this is quite a small unit the kilonewton (kN) will be encountered more frequently.

Pressure

When a force acts uniformly over an area instead of at a point, it is said to be exerting a **pressure**. Pressure is measured in newtons per square metre (N/m^2). This, too, is a very small unit, and a kilonewton per square metre (kN/m^2) is a more practical unit. However, as an alternative to this the **bar** is being used. One bar is equivalent to 100 000 N/m^2, and is also roughly equivalent to the pressure normally exerted by the atmosphere. Thus, an inflation pressure of 80 kN/m^2 (alternatively represented as 0.8 bar) might be recommended for a rear tyre of a tractor. This means that every square metre of the inside surface of the tube would be subjected to a uniform force of 80 kN.

A pascal (Pa) is occasionally used as an alternative for 1 N/m^2.

Moments

When a force exerts a turning effect, as when pulling or pushing at the end of a spanner to turn a nut, the force on the spanner is said to be exerting a **turning moment**, or **torque**. The torque is calculated by multiplying the force by the distance measured from the turning centre perpendicular to the direction of the force. Torque is measured in newton-metres (Nm) (Fig. 2.1). The torque developed by an engine is often quoted and this refers to the turning or twisting effect being given by the crankshaft.

Cylinder-head nuts and other nuts are tightened to a given torque, and a special torque spanner or wrench can be used to tighten such nuts. When the nut has been tightened to the correct torque a ratchet in the wrench handle releases, making it impossible to tighten the nut further.

Work

When a force moves it is said to have done **mechanical work**. Thus, if a worker raises a box of plants to waist level, he has exerted a force, and the force has moved, so work has been done. If he continues to hold the box at waist level he does no further work, even though holding the box may require a considerable effort. Thus, if a force of

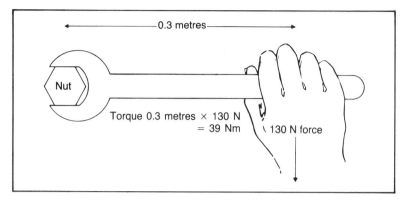

Fig. 2.1 Applying torque to nut.

200 N has to be used to raise the box, and it is lifted 0.9 m, the work done is 200 × 0.9 = 180 Nm.

Energy

The ability to do mechanical work as defined in the previous section is referred to as **energy**.

There are many different forms of energy. For instance, chemical energy, as in fuels, which can be released in an engine to do work; heat energy which can be utilized in a steam-engine to do work; electrical energy which when released from a battery can operate a starter motor; mechanical energy possessed by a moving object, or by an object suspended so that energy is released and work done when it falls; and there are a number of others.

When energy in one form is used it reappears as energy in another form. For instance, when fuel is burnt in an engine releasing chemical energy some is changed into useful mechanical energy. The remainder, resulting largely from friction, is changed into heat energy which is wasted.

The amount of energy, whatever its form, is measured in joules (J). For instance, heat outputs from boilers used to heat glasshouses are indicated in megajoules (MJ).

Power

Power is defined as the 'rate of doing work'. The watt (W) is the unit of power, and 1 W is equivalent to 1 joule per second (J/s), or 1 newton-metre per second (Nm/s).

Thus, if a tractor pulling a plough with a force of 18 kN travels a distance of 2 m in one second, the rate of doing work, which is the power the tractor is exerting, is

18 000 × 2 = 36 000 Nm/s

This is equivalent to 36 000 W or, in more practical units, 36 kW.

Brake power is the power developed at the crankshaft of an engine, whereas drawbar power is the power developed at the drawbar, and power-take-off power is the power developed at the power take-off. These terms are widely used in information relating to horticultural tractors.

Rated power may also be referred to. This is the power that the tractor engine is designed to produce continuously, hour after hour. It is usually the engine power at slightly less than maximum engine speed.

Questions

1. What is the prefix for × 1 000 000?
2. What is the abbreviation for cubic metres?
3. What is a force?
4. Give three practical examples in horticulture of the action of a force.
5. What is pressure?
6. What is the moment of a force of 50 kN acting at a distance of 2 m measured perpendicularly from the direction of the force to the turning centre?
7. Give three examples of different forms of energy.
8. What is power?

3 Electricity

Conductors and non-conductors

Electricity will flow through conducting materials but not through non-conductors. Conducting materials include metals, some non-metals such as carbon, and water provided that it contains some slight impurity. Some materials are better conductors than others, and copper, which is an excellent conductor, is used for most electrical work although other materials are used for special purposes. For instance, a carbon brush is frequently used where an electrical current has to flow to a moving component because of its self-lubricating properties.

Circuits

Electricity is made to flow through a conductor by 'electrical pressure' from a battery or a generator. An electrical socket is also an indirect source of electrical pressure as it is ultimately connected to the generator at the power-station. But a current will only flow if there is an unbroken circuit of conductors from the source of electrical pressure, through the electrical appliance or equipment, and back again. Any switch or break in the circuit, or the incorporation of a non-conductor, will prevent a current flowing. (Fig. 3.1).

In many ways the flow of electricity in a conductor can be compared with a water system. In each case the flow of electricity or water through a conductor or pipe depends on the pressure, the flow rising as the pressure rises, and vice versa. Also, in each system there is resistance to flow, and a decrease in resistance will increase the flow.

Electrical units

The size of a current is measured in amperes, usually abbreviated to amps (A). Electrical pressure and resistance are measured in volts (V) and ohms respectively. These three quantities are related by Ohm's law which indicates that the current flowing through a conductor is equal to the electrical pressure divided by the resistance of the conductor. That is

$$\text{Current} = \frac{\text{Electrical pressure}}{\text{Resistance}}$$

Note that if the pressure is increased or the resistance decreases, the current increases, as indicated earlier.

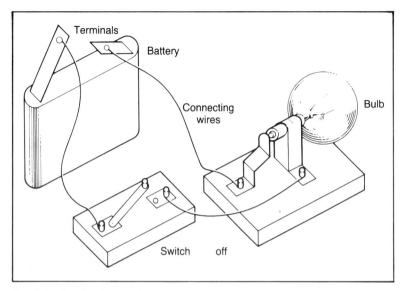

Fig. 3.1 Simple electrical circuit.

Electrical power

Electrical power is measured in watts. The power requirement of a simple appliance can be calculated by multiplying the current (in amps) flowing through it by the electrical pressure (in volts) causing the current to flow. For instance, an electric heater connected to a 250 V supply, with a current of 4 A flowing through it, has a power requirement of 250 × 4 = 1 000 W.

The practical unit of power is the kilowatt which is equivalent to 1 000 W. When 1 kW of electrical power is used continuously for an hour, 1 kilowatt-hour (kWh) of power has been consumed. This is the commercial unit of electricity which is used as the basis of tariffs charged by Electricity Boards. Meters installed by Electricity Boards are fitted with dials or figures showing numbers of units. Power consumption in units over a period of time is calculated by subtracting the reading at the beginning of the period from the reading at the end of the period.

A.C. supplies

In this country electrical sockets provide a pressure of 200 to 250 V, but it is an alternating current (a.c.) supply. An a.c. supply does not provide a steady current, but a current which fluctuates both in size and direction. The changes follow a regular pattern or 'cycle', and 50 complete cycles occur every second. For simple practical purposes the regular changes in the supply can be ignored as they have no noticeable effect on lights and heaters, although they do make possible the operation of transformers and electric motors.

Supply cables

The supply of current for an electrical load such as a heater or motor is carried by two conductors which are insulated from each other and are known as the live and neutral conductors. Cables carrying such a supply will probably contain a third, insulated conductor, the earth, within the outer protective sheath of rubber or plastic.

To identify the individual conductors different-coloured insulation is used on each. Appliances sold in this country are wired in accordance with an internationally-agreed code which uses brown for the live, light blue for the neutral and green and yellow stripes for the earth. These combinations tend to be easily distinguished in poor light and by those who are colour-blind. (Fig. 3.2).

Prior to 1970 in this country, red was used for the live, black for the neutral and green for the earth. Old equipment using red-, black- and green-coloured wires may still be in use. Great care must be taken when fitting plugs or renewing flexes on such appliances.

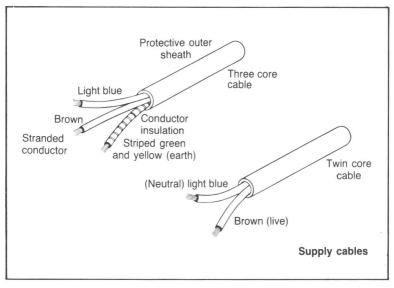

Fig. 3.2 Supply cables.

Single and three phase

The normal supply to a house and for many purposes in horticulture, such as lighting and heating, is referred to as single phase. When high demands are made on the supply, particularly when large electric motors are to be used, a three-phase supply may be required. The three-phase supply to a motor is carried by three conductors without any standardized colour coding and, in addition, there will be an earth wire. Three-phase motors are usually smaller, cheaper, and have more acceptable starting characteristics than their single-phase equivalents.

Earthing

The provision of an earth conductor is a safety measure. The metal frame or casing of an electric motor or other electrical appliance is connected to earth. Thus, if there should be a fault in the equipment or a failure in the insulation which would make the exposed metal parts live and dangerous, the current flowing in them is earthed. Only a slight tingle at most would be felt by anyone touching the metal. It is important that any appliances with exposed metal parts should be earthed, and they should not be used on any supply without an earth, such as a two-pin plastic light socket.

The main earth conductor from a building may have an earth-leak trip device incorporated in it. If too large a current flows through the earth conductor as a result of a fault in the electrical system, the earth-leak trip switches off the main electricity supply. Pressing a button or switch on the earth-leak trip will restore the supply, but if it trips again there is a definite fault which must be traced and rectified.

Some portable tools are 'double insulated' and require no earth wire. Their outer covering is made from an insulating material which provides protection for the user in case of electrical fault. Double-insulated appliances carry the identification mark shown in Fig. 3.3.

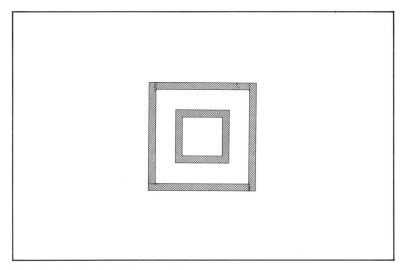

Fig. 3.3 Double insulation symbol.

Fuses

To prevent overloading and damage with the possible risk of fire caused by too large a current passing through a conductor of an electrical appliance, a fuse is fitted into the 'live' side of the circuit. For most circuits these consist of soft lead–tin alloy wires which melt if too large a current is allowed to pass through them so that the circuit is broken. Fuses are often manufactured in the form of cartridges which are

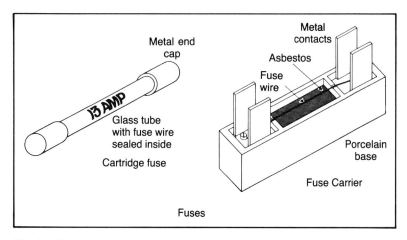

Fig. 3.4 Fuses.

labelled and colour coded according to the size of current that they are intended to carry. (Fig. 3.4).

Replacing a 'blown' fuse with one of a higher current rating is dangerous. A repeated failure of a fuse of the correct rating indicates a fault in the circuit which must be located and repaired. Fuses for a number of circuits may be grouped together in a single fuse box, but in some circuits are fitted in the plug.

Wiring a plug

When wiring a three-pin plug, great care must be taken to attach the wires to the correct terminals according to the colour coding. The live, neutral, and earth terminals are usually indicated by the letters, L, N, and E. The conductors should be tightly attached to the terminals, care being taken when stripping the insulation not to damage or break any of the fine strands which make up the conductor wire and give it flexibility. The cable leading to the appliance should be firmly gripped by the cable clamp to prevent any strain on the cable from pulling the conductors off the terminals. Individual conductor wires should not be visible outside the plug (Fig. 3.5)

The terminals of two-pin plugs often are not labelled, and the live or neutral conductors can be attached to either terminal. Only lights or appliances that do not require an earth can be connected to two-pin plugs without provision for an earth.

Safety

No wiring or repairs to an electrical system or appliance should be attempted without first switching off at the mains and removing the fuse for that circuit. If someone has received an electric shock, he should not be touched unless the power has been switched off, or until he has been pushed away from the electrical supply with a non-conductor such as

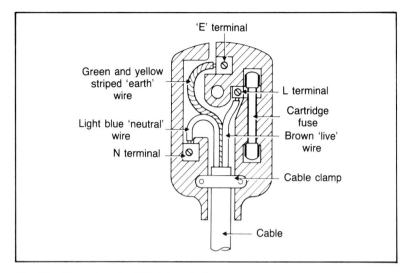

Fig. 3.5 **Correctly wired 13 A fused plug.**

a length of dry wood. Medical help should be then summoned and artificial respiration given if necessary.

Questions

1. What is the difference between a conductor and a non-conductor?
2. Give three examples of good conductors.
3. Give three examples of good insulators.
4. What are the units of electrical current, pressure, and resistance?
5. What is the relationship between electrical current, pressure, and resistance?
6. What are the insulation colours for live, neutral, and earth cables under the present international colour-coding system?
7. What is the purpose of a fuse in a circuit?
8. What is the importance of having a good earth connection to electrical appliances?

4 An introduction to types of tractors used in horticulture

Horticultural tractors can be defined in general terms as machines capable of pulling, carrying or operating a variety of horticultural implements and machines.

They can be divided into two classes: tractors on which the operator rides, and pedestrian-operated machines behind, or beside which, the operator walks as he controls its work.

Ride-on tractors

Ride-on tractors may have three or four wheels, the three-wheeled version having a tricycle layout, with the single wheel fitted either at the front or the rear and normally used for steering (Fig. 4.1).

Fig. 4.1 A specialized small tractor with hydrostatic drive.
By permission of Nickerson Turfmaster Ltd.

Fig. 4.2 Medium-sized tractor working in hops.
By permission of Massey Ferguson (UK) Ltd.

Conventional wheeled tractors range in size from tiny but effective models of 8 kW or less to giants of 260 kW or more, seldom used in horticulture. In between there are a variety of models of different layout and specification. Many of these were designed for agriculture and may not be ideally suited for horticultural work. Special adaptations for horticulture such as very low orchard models, very narrow vineyard models and row-crop models with a big vertical clearance underneath them have been available for many years (Fig. 4.2).

Small tractors are designed with horticulture very much in mind. They are used wherever power requirements are low and in confined areas such as glasshouses, and can carry out a wide range of duties. Many have two-wheel drive (Fig. 4.3), but in others the steering-wheels are driven as well to provide four-wheel drive which affords much better wheel grip (Fig. 4.4). In a few cases four-wheel-drive tractors are steered by articulating them at the point where engine and gearbox meet.

Tractors in the 26 to 48 kW range are commonly used by local authorities for a wide range of work, but as many of them spend a high proportion of their working time travelling on the public highway a modified industrial version is often used.

Tractors weighing more than 559 kg are fitted with strong safety cabs or frames intended to protect the driver if they should overturn and are designed so that the noise level at the driver's ear does not exceed 90 decibels.

Alternatively, riding tractors may be fitted with tracks which provide better traction than wheels and exert lower ground pressure. Track-layers are very manoeuvrable and can be slewed around very sharply, although this marks the surface of the ground very noticeably. They tend to be slow and are not permitted on the road without pads fitted on the track plates to protect the road surface. They are rarely used in

Fig. 4.3 Two-wheel-drive small tractor with hydraulic-drive cylinder mowers.
By permission of Lely Iseki Tractors.

Fig. 4.4 Four-wheel-drive small tractor rotovating.
By permission of the Ford Motor Co. Ltd.

Fig. 4.5 Track-layer subsoiling.
By permission of Massey Ferguson (UK) Ltd.

horticulture except as large units for pulling drainage equipment, and earth-moving equipment used in landscaping (Fig. 4.5).

Pedestrian-operated tractors

Pedestrian-operated tractors are nearly always purpose designed for horticultural work. They have either two wheels or have to be balanced on one. Although many are designed to carry or pull a range of implements, others are made as single-purpose units with equipment such as rotary cultivators permanently fitted or so difficult to remove that they are seldom used for other purposes (Fig. 4.6).

The handlebars on pedestrian-controlled machines are normally adjustable in height to suit different operators. They may be offset to either side so that the operator can walk to one side when it is undesirable or uncomfortable for him to walk behind the machine as, for instance, when rotovating.

Even pedestrian-operated machines can occasionally be fitted with seats mounted on an extra pair of wheels, or rollers, towed behind so that the operator can ride.

General construction

Whatever the type of tractor there are a number of major units common

Fig.4.6 Pedestrian-controlled rotary cultivator.
By permission of the Howard Rotavator Co. Ltd.

to them all which can be recognized, although they may lie in slightly different relative positions. These major units include the engine, the clutch, the gearbox, and the differential through which the engine power is transmitted to the wheels; and on bigger tractors the hydraulic lift system. The purpose and operation of all these units will be explained in later chapters.

Questions

1. What are the three adaptations of the four-wheel general-purpose tractor made specially for horticultural purposes?
2. What other type of four-wheel tractor is designed and built solely for horticultural use?
3. What are the advantages and disadvantages of tractors fitted with tracks?

5 The working principles and basic construction of internal combustion engines

The basic working principles of all internal-combustion engines used as horticultural power units are similar. A mixture of a small quantity of fuel with the correct amount of air is exploded in a cylinder which is closed at one end. As a result of the explosion heat is released and this causes the pressure of the burning gases to increase. This pressure increase forces a close-fitting piston to move down the cylinder, and this movement is transmitted to a crankshaft by a connecting-rod so that the crankshaft turns through half a revolution.

To obtain continuous rotation of the crankshaft this explosion has to be repeated. Before this can happen the used gases have to be expelled from the cylinder, a fresh charge of fuel and air admitted, and the piston moved back to its starting position (Fig. 5.1).

This sequence of events is known as a working cycle, and all internal-combustion engines used in horticulture operate on either a two-stroke cycle, or a four-stroke cycle. Both petrol engines and diesel engines are

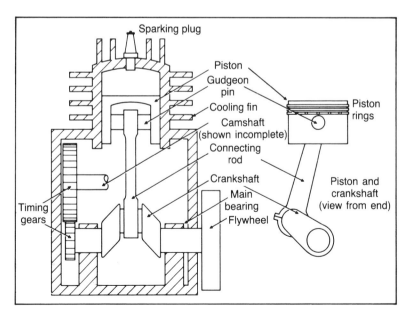

Fig. 5.1 A section through an engine showing the main components.

used. The stages in the two working cycles for both types of engine in the sequence which they occur are as described below.

Petrol engines

The four-stroke working cycle – petrol engines

First stroke – the induction stroke

The piston is moved down the cylinder and a fresh charge of mixture (air plus petrol vapour) is drawn in through an induction pipe which leads to an open inlet valve in the top of the engine cylinder (Fig. 5.2).

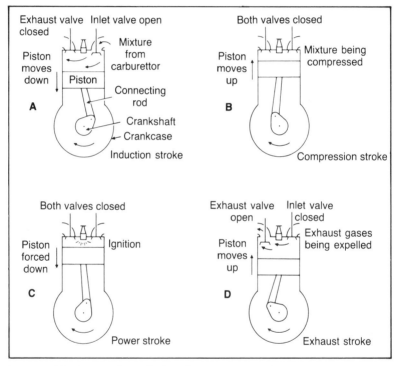

Fig. 5.2 The four-stroke cycle (spark-ignition engine).

Second stroke – the compression stroke

The inlet valve is closed, and the piston is moved up the cylinder, compressing the new charge of mixture. This is ignited just before the end of the stroke by a spark produced in the cylinder by an electrical circuit (described in Ch. 8).

Third stroke – the power stroke

The rise in pressure in the cylinder caused by the burning gases forces the piston to move down the cylinder.

Fourth stroke – the exhaust stroke

During the final stroke of the cycle the piston is moved up the cylinder to expel the burnt gases through an open exhaust valve in the top of the cylinder into the exhaust pipe which directs them away into the atmosphere. Only during this stroke is the exhaust valve open.

As a stroke is defined as the movement of the piston from one end of the cylinder to the other and results in half a turn of the crankshaft, the whole cycle is completed in four strokes and two complete revolutions of the crankshaft.

The two-stroke working cycle – petrol engines

In this cycle the whole sequence of events are compressed into two strokes and one complete revolution of the crankshaft. There are no valves and the gas movements take place through holes called ports in the cylinder walls. The crankcase in which the crankshaft rotates is gas tight and is involved in the cycle.

First stroke – the induction/compression stroke

The piston is moved up the cylinder and as it does so it covers and closes two of the ports, the exhaust port and the transfer port which are normally almost opposite each other. This traps a charge of fresh mixture in the cylinder, and further upward movement of the piston compresses this.

Further movement of the piston also uncovers a third port lower down

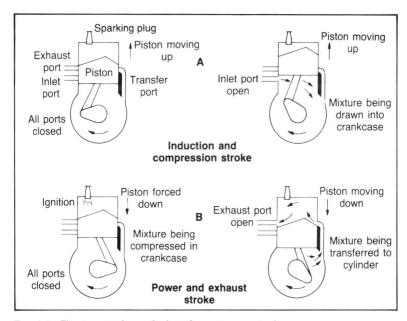

Fig. 5.3   The two-stroke cycle (spark-ignition engine).

in the cylinder which is called the induction port, and more fresh mixture is drawn through this into the crankcase.

Just before the end of this stroke the mixture in the cylinder is ignited as in the four-stroke cycle (Fig. 5.3).

Second stroke – the power/exhaust stroke

The rise in pressure in the cylinder caused by the gases burning forces the piston to move down the cylinder. Part way through the stroke the piston covers and closes the induction port, trapping the mixture drawn into the crankcase during the previous stroke and then compressing it. Further downward movement of the piston uncovers first the exhaust port, and then the transfer port, allowing the burnt gases to flow out through the exhaust pipe and the fresh mixture under pressure in the crankcase to transfer into the cylinder. Frequently a shaped piston crown deflects the incoming mixture up and across the top of the cylinder so that it helps to scavenge the exhaust gases.

Although this cycle provides one power stroke per revolution it is not necessarily more efficient than the four-stroke.

Working cycles for diesel engines

Diesel engines are also used to power horticultural machines. These operate on the same working cycle as petrol engines, but air on its own is drawn into the cylinder instead of a mixture of air and fuel in vapour form. Compression of the air during the compression stroke makes it very hot, and towards the end of the stroke diesel fuel is injected into the cylinder in the form of minute droplets which ignite spontaneously, causing the rise in pressure which forces the piston down. No electrical ignition system to provide a spark is necessary.

Engine features

In both two-stroke and four-stroke working cycles only one stroke is a power stroke, and the engine crankshaft has to be kept turning until the next power stroke occurs. The moving parts are given sufficient momentum for this to happen by a heavy flywheel firmly fixed on a taper at one end of the crankshaft. This stores energy during the power stroke and releases it again during the other strokes in the cycle, thus keeping the engine speed steady.

The valves in an engine working on the four-stroke cycle are known as poppet valves. Each one consists of a mushroom-shaped head mounted on a long thin stem. When closed, a valve spring holds the valve head down on to a seat to give a gas-tight seal. The valve is opened by a cam which, as it turns, forces the valve off its seat against the pressure of the valve spring (Fig. 5.5). All the cams are mounted on a common camshaft which is driven from the engine crankshaft by timing gears. These gears have timing marks on them which are provided so that the camshaft drive can be reassembled correctly after an overhaul to ensure that the valves will open and close at the correct point in the cycle.

Side-valve engines have valves which open upwards into an extension of the cylinder-head space and are directly operated by the

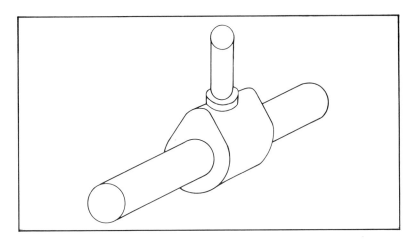

Fig. 5.4 Cam and follower.

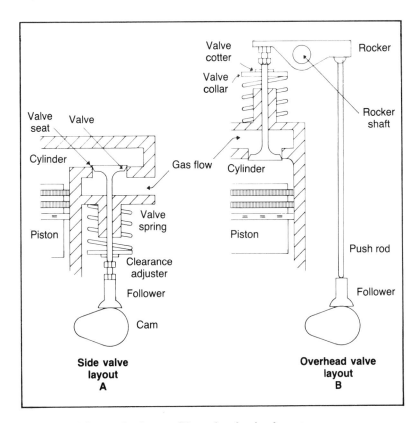

Fig. 5.5 (A) Side-valve layout; (B) overhead-valve layout.

camshaft. Overhead valves, used in more complex engines, are valves which open downwards into the engine cylinder and are operated by a complicated arrangement of push rods and rockers (Fig. 5.5).

Single-cylinder engines are used to drive small machines such as pedestrian-operated mowers. Three-cylinder and four-cylinder engines are the most common arrangement in tractors used in horticulture.

Questions

1. What is the basic working principle of an internal-combustion engine which forces the piston along the cylinder?
2. What is the sequence of working strokes during the four-stroke cycle in both petrol and diesel engines?
3. What happens during each stroke?
4. What happens both above and below the piston during each of the working strokes of the two-stroke cycle for both petrol and diesel engines?
5. What are the differences between the working cycles of diesel engines and petrol engines?
6. What is the difference between an overhead valve and a side valve?

6 Lubrication and cooling systems

Although the operating principles of an internal-combustion engine were described in Chapter 5, the engine cannot function continuously without added refinements.

Lubrication systems

Whenever two surfaces rub together friction occurs. In an engine some of the fuel energy which could otherwise be converted into useful power has to be used to overcome friction and results in heat being produced at the bearing surfaces. Rapid wear may also result. High engine speeds and high bearing loads due to an engine slogging or being overloaded make the problem more acute.

The introduction of a lubricant between the surfaces in contact reduces friction and acts as a coolant. In addition it flushes away metal filings and dirt which might otherwise score the smooth and highly polished bearing surfaces.

A supply of oil to all the important engine bearings and wearing surfaces is provided by a lubrication system. In a small engine this will probably be a simple splash-feed system. In a large multi-cylinder tractor engine it will be a more complex force-feed system.

Splash-feed systems

In a splash-feed system oil from the crankcase of the engine, which serves as a reservoir, is flung or splashed on to the timing gears and the cylinder walls. This is done either by a dipper mounted on the big-end cap which dips into a narrow trough within the sump containing oil each time that the crank rotates, or by a paddle wheel driven by the timing gears. This oil lubicates the cylinder walls. Some passes through holes in the piston walls to lubricate the gudgeon pin, while surplus oil runs down the connecting-rod and passes through a hole to lubricate the big-end bearing. Some of the oil running back down the cylinder walls is collected and directed to the main bearings which support the crankshaft before returning to the reservoir for further circulation (Fig. 6.1).

Force-feed systems

For more complicated and heavy-duty engines a more regular supply of oil under pressure is necessary. To provide this a pump, which is normally mounted in the crankcase, is used. It is driven from the timing gears or from the camshaft. The pump forces oil from the oil reservoir

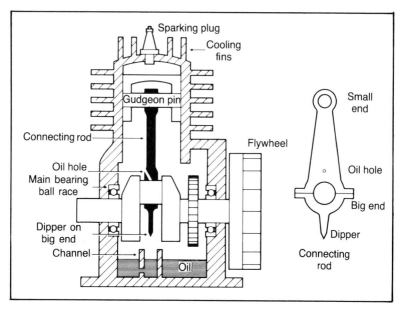

Fig. 6.1 Splash-feed lubrication system.

in the crankcase through a pipe or drilling in the engine block. These are known as main oil galleries, and branches lead off to each of the main bearings and to a jet spraying oil on to the timing gears. A further extension may supply a controlled amount to the rocker-shaft bearings. The oil from the main bearings continues under pressure through drillings in the rotating crankshaft to lubricate the big-end bearings. As in a splash-feed system, oil expelled from the big ends as they rotate is flung up on to the cylinder walls, lubricating these and the gudgeon pin before draining back into the sump (Fig. 6.2).

Coarse particles of dirt in the oil are excluded from the pump by a coarse gauze screen over the pump inlet. All the oil leaving the pump is passed through a full-flow filter where finer particles are removed before it is passed on through the drillings to the easily damaged bearing surfaces.

As the pump is driven by the engine, pressure varies with engine speed and may be observed to be quite low at idling speeds. Maximum pressure is controlled by a pressure-release valve which opens at a preset pressure and allows some of the oil to return to the sump without passing through the complete system.

Low oil pressure is an indication of worn parts, low oil level, or the incorrect grade of oil in the sump.

Lubricating oils

Engine oils of different thickness or viscosity are available. Thick oils tend to give better bearing protection, but make it harder to turn over the engine at cold starts and flow only slowly round the engine when

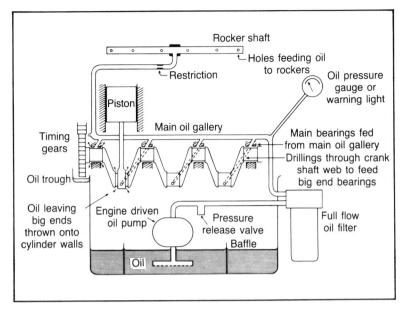

Fig. 6.2 Force-feed lubrication system.

it is cold. Ordinary oils tend to become thinner when hot, giving reduced protection. Multigrade oils can be obtained which thin only slightly as their temperatures rise.

Oil thickness is indicated by an SAE (Society of Automative Engineers) scale, smaller numbers indicating the thinner oils. Thus an SAE 10 is a thin engine oil and an SAE 50 is a thicker one. SAE 90 and 140 are transmission oils, and SAE 10/30 is a multigrade engine oil.

Extreme-pressure (EP) oils contain additives which make them suitable for use in heavily loaded gearboxes.

Heavy-duty (HD) oils should always be used in diesel engines. They too are fortified with additives and have detergent properties. This means that carbon and other impurities are kept in suspension in the oil and not allowed to settle out where they might cause blockage of an oilway.

Universal oils are available which are formulated to provide properties which make them suitable for use in all the main units of the tractor, engine, gearbox, and hydraulics.

Maintenance of lubrication systems

After an engine has been in use for some time and becomes worn, or if leaks occur, the oil level in the engine may drop to a point where there is insufficient oil left in the sump to lubricate the engine safely. A dipstick is fitted in the sump so that the engine oil level can be checked (Fig. 6.3). This should be done daily, and if the level has dropped below the 'full' mark on the dipstick new oil should be poured into the sump until the correct level is reached. Overfilling should be

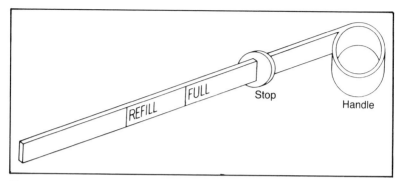

Fig. 6.3 A dipstick.

avoided. The dipstick check is only accurate when the engine is level.

Oils become dirty and deteriorate chemically in use. Thus regular oil changes carried out at the intervals specified in the manufacturer's instruction book are necessary. If the engine has a force-feed system a replacement filter cartridge or complete new filter will have to be fitted.

The lubrication system maintenance specified by the manufacturer must not be neglected or the engine may wear rapidly. It is essential that a record of use is kept for every machine equipped with an engine so that it is known when oil and filter changes are necessary. Changes in the appearance or 'feel' of an oil are not reliable indications of the need for an oil change. Fresh oil soon goes black in older engines.

Two-stroke lubrication

In a spark-ignition engine operating on a two-stroke cycle, transfer of the mixture takes place through the sump. Thus no oil is used in the sump of this type of engine and lubrication is provided by oil which is mixed with the fuel. Two-stroke mixture which already contains the correct amount of oil can be purchased, but it can be prepared by adding two-stroke oil, which is formulated to mix readily with petrol, in the correct proportion. These proportions vary quite widely between 16 to 1 and 50 to 1 (petrol to oil respectively) for different engines, and the recommendations of the manufacturer should always be followed.

Cooling systems

When fuel is burnt in an engine cylinder all the heat produced is not used usefully in raising the pressure in the cylinder. Some is wasted in heating up the metal of the cylinder walls and cylinder head. A cooling system is necessary to remove this surplus heat which otherwise causes rapid temperature increases leading to expansion of the moving parts which may take place at different rates, and possibly to seizure of the piston in the cylinder.

Two different cooling systems are used. Small engines are usually air-cooled, whereas larger multi-cylinder engines are most frequently liquid-cooled.

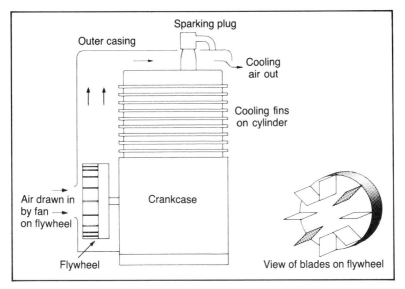

Fig. 6.4   Air-cooling system for single-cylinder engine.

Air-cooling systems

The cylinder of an air-cooled engine has fins protruding from it to increase the area of metal from which heat transfer can take place. It is normally enclosed in a sheet metal casing called a cowling. The externally-mounted engine flywheel has blades projecting from its face so that it acts like a fan, drawing air in through a hole in the cowling in front of the flywheel and directing it around the finned cylinder to an outlet vent. Little control of engine temperature is usually possible (Fig. 6.4).

Maintenance of air-cooling systems

Maintenance is confined to keeping all the airways free from blockage by dirt or other materials, such as grass, which otherwise will limit air flow and cause overheating. This is done by removing the cowling and cleaning out the dirt with a stiff brush or compressed air. Occasionally a separate fan which is belt-driven from the crankshaft provides the air flow, and the belt tension will have to be checked and adjusted if necessary.

Liquid-cooling systems

In liquid-cooled systems the surplus heat is removed by a liquid, normally water, which flows through passages in the cylinder head and block. An impeller pump, which is belt-driven from the crankshaft, circulates the water. The water leaves through a header pipe from the top of the engine to a radiator where it is cooled before returning to the lower part of the block through a second pipe (Fig. 6.5).

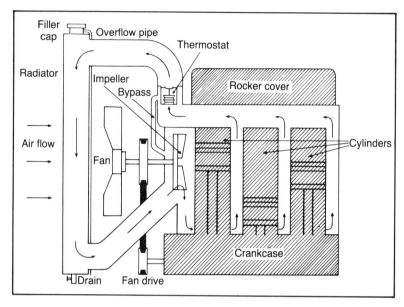

Fig. 6.5 Diagram of a liquid-cooling system.

The radiator is made up of a number of narrow, vertical pipes through which the water flows. A fan mounted on the pump shaft draws air between these pipes cooling the water in them.

Thermostat

A temperature-operated valve, called a thermostat (Fig. 6.6), fitted in the header pipe controls the engine temperature. Until the working temperature is reached the thermostat is closed and the water is short-circuited back into the block and circulated around it without passing through the radiator. Once the engine has reached its working temperature the thermostat opens and the water makes a full circuit through the radiator and is cooled. The thermostat allows the engine to maintain a more even running temperature and assists it to warm up more quickly, reducing wear.

As the water temperature in the system rises the wax expands and partially expels the central pin, opening the thermostat. As the water cools the reverse happens and closing is assisted by a spring.

Pressure cap

A pressure cap is fitted on the radiator of most modern engines. This cap stays closed until the pressure in the cooling system has risen well above atmospheric pressure. Because the boiling-point of water rises with increase in pressure, fitting a pressure cap allows the engine to be run at a higher working temperature, and thus more efficiently without fear of the water boiling. The pressure at which a cap opens is marked on it and may be as high as 1 bar.

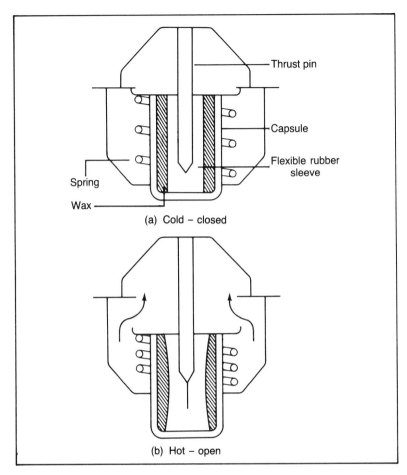

Fig. 6.6 Section through a wax-type thermostat.
(a) Closed; (b) open.

Maintenance of liquid-cooling systems

The level of the water in the radiator should be checked daily. If the level has dropped appreciably the radiator should be topped-up with clean water. If the cooling system contains antifreeze the topping-up should be done with antifreeze solution.

Antifreeze must be added to the cooling water to prevent freezing, which can occur even when the engine is running if the air temperature is below freezing-point. A 25 per cent solution of antifreeze and water, that is 3 parts of water to 1 part by volume of antifreeze, gives adequate protection for most British conditions; although a 33 per cent solution, 2 parts of water to 1 part of antifreeze, is now commonly recommended by manufacturers.

The system should be thoroughly flushed out before antifreeze is put in early in the autumn, and again after draining in the spring, to remove dirt, sludge and flakes of rust. A good brand of antifreeze will consist of ethylene glycol and contains a corrosion inhibitor to prevent rusting.

Some antifreezes with this formulation can be left in for the whole summer, and even for a second year, provided that topping up is done with antifreeze mixture of the correct strength so that the solution in the system is not weakened.

Fan-belt tension has to be adjusted so that there is approximately 13 mm (0.5 in) of sideways play on the greatest length of the belt.

The hose clips should be checked for tightness and the whole system checked for leaks, particularly when antifreeze is put in as this will leak out more easily than plain water.

Questions

1. What are the harmful effects on an engine caused by friction?
2. How are all the moving parts of a small engine lubricated by a splash-feed system?
3. What is the purpose of a filter in a force-feed lubrication system?
4. What are multigrade oils?
5. Why has oil in an engine to be changed at regular intervals?
6. Why is an engine-cooling system necessary?
7. What maintenance does an air-cooling system require?
8. What is the purpose of a thermostat in a liquid-cooling system?
9. What quantity of antifreeze should be added to a cooling system containing four litres to prevent freezing in normal conditions?
10. How should the fan belt be tensioned?

7 Fuel systems

In Chapter 5, it was stated that the fuel and air necessary for combustion in a petrol engine are supplied as a mixture with the fuel in vapour form, whereas fuel and air for a diesel engine are supplied separately, the fuel being injected into the cylinder as a mist of minute droplets. In each case it is the task of the fuel system to supply clean fuel and air in the correct proportions, in the correct physical form, and at the correct point in the cycle.

The air cleaner

Air cleaners are fitted to both types of engine. In normal operations a petrol engine draws in approximately 10 000 litres of air for every litre of fuel it uses. Engines in use in horticulture frequently have to work in dusty conditions, but even when working in relatively clean conditions the large volume of air being drawn into the engine results in an appreciable quantity of gritty dust being drawn in as well. If allowed to reach the engine cylinders this will result in worn and damaged valve seats, piston rings, and cylinder walls, with subsequent loss in engine power and efficiency.

Simple filters

The simplest type of air cleaner used on small horticultural engines takes the form of a fine gauze or felt sandwiched between two support gauzes fitted over the mouth of the air induction pipe which is enlarged to provide a large filtration area (Fig. 7.1) A plastic sponge often replaces the felt (Fig. 7.2).

Maintenance

Blockage of the air filter by dirt will reduce the amount of air drawn into the engine and so it is essential that the filter is cleaned regularly. The instruction book for the engine will indicate how frequently this should be done under normal conditions. When working in unusually dirty conditions the maintenance should be carried out more frequently.

Cleaning is done by removing the filter element and washing it thoroughly in paraffin, or warm water and detergent. After draining it can be replaced. When a sponge is used for filtration it should be lightly oiled with engine oil after cleaning, and the surplus squeezed out, before the sponge is replaced in its container.

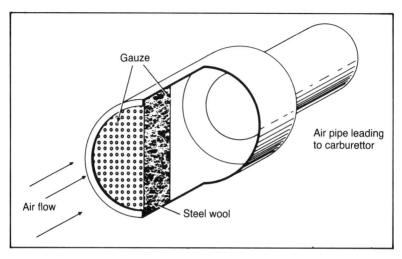

Fig. 7.1 Part section through gauze air cleaner.

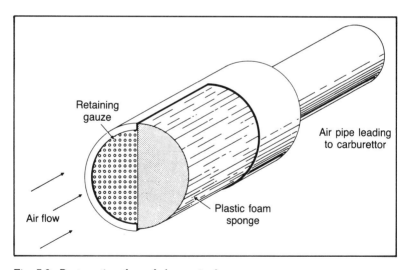

Fig. 7.2 Part section through foam air cleaner.

Two-stage cleaners

More complicated two-stage air cleaners are used for some engines, including tractor engines. Larger particles of dust are removed by a centrifugal pre-cleaner and finer dirt is removed by an oil-bath cleaner (Fig. 7.3).

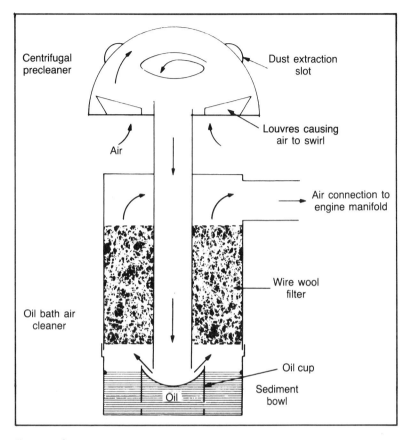

Centrifugal precleaner

Dust extraction slot

Louvres causing air to swirl

Air

Air connection to engine manifold

Wire wool filter

Oil bath air cleaner

Oil cup

Oil

Sediment bowl

Fig. 7.3 Section through a pre-cleaner and oil-bath air cleaner.

Maintenance

The oil level should be topped-up if necessary daily, and the oil should be changed every 50 hours or when more than 6 mm of dirt has accumulated in the bowl.

Dry air cleaners

A dry air cleaner is often an alternative to an oil-bath type. In these the air passes inwards through a circular filter made of paper. To provide the large surface necessary for filtration the paper is folded many times concertina-fashion. Tractor air cleaners of this type may have a second cylindrical paper filter inside the first through which the air also has to pass. It can be fitted with a warning light operated by a pressure switch which lights up when there is a pressure difference across the filter, indicating that it is becoming blocked and requires maintenance.

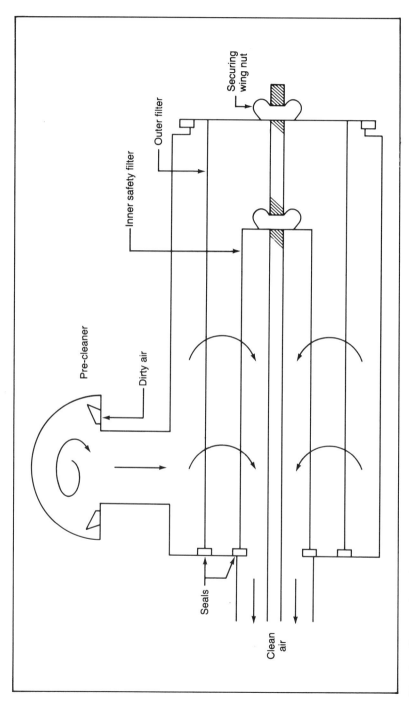

Fig. 7.4 Section through a two-stage dry air cleaner.

Maintenance

A dry air cleaner can be cleaned by tapping it on a flat surface to knock off the dust. If it is very dirty it can often be washed in warm water and detergent, flushed with clean water, and then air dried. The use of an air line is not recommended as the air pressure can easily damage the filter. After washing a number of times, or after a year of use, the filter should be replaced. Dry filters should never be oiled.

Fuel systems for petrol engines

For perfect combustion a mixture of 14.7 parts of air to 1 part of petrol by weight is required. A richer mixture, that is, a mixture containing a greater proportion of fuel, will provide maximum power from the engine but at the expense of higher fuel consumption. However, weaker mixtures, up to 18 parts of air to 1 part of petrol, improve fuel economy.

It is the job of the carburettor to provide the required mixture at all engine speeds and loadings, and to ensure that the fuel is completely vaporized and properly mixed with the air.

Working principles of a carburettor

The basic principle of all carburettor designs is that when air flows over the end of a narrow tube or jet containing liquid, a drop in pressure occurs which results in some liquid being drawn out into the air stream. The same principle is used in the construction of hand plant sprayers and scent sprays (Fig. 7.5).

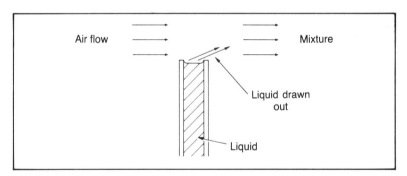

Fig. 7.5 Effect at a jet.

The quantity of liquid drawn into the air stream increases as the speed of the air flow over the jet rises, and also is greater if the jet is made larger.

In practice, the fuel level in the jet is maintained by a float chamber. The fuel level in the jet and in the float chamber are always the same. As fuel is used the level in the float chamber sinks, the float in the float chamber sinks with it and the needle valve comes off its seat allowing more fuel into the chamber from the fuel tank. When the fuel level has risen to its correct level the float presses the needle valve back on to its seat and cuts off the fuel flow.

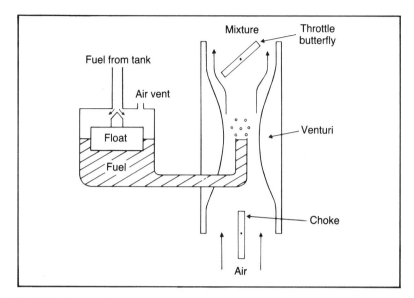

Fig. 7.6 Section through a simple carburettor.

The velocity of the air flowing over the jet is increased by a constriction in the induction pipe known as the venturi. A throttle butterfly valve provides an adjustable obstruction in the induction pipe which can be used to control the flow of mixture to the engine, and thus the engine speed. As the butterfly valve is turned into the 'accelerate' position the air flow over the jet increases and more fuel is drawn out into the air stream, keeping the mixture strength constant (Fig. 7.6). Unfortunately, the mixture strength given by this simple carburettor will not be constant over the engine speed range, and in more advanced designs compensating devices are fitted to ensure that the mixture strength does not change.

A second buttterfly valve, the choke, is used to provide a richer mixture for cold starts when it is too cold for all the fuel drawn into the engine to vaporize.

A second jet, which is often adjustable, is fitted in the induction pipe near the throttle butterfly and comes into use when the engine is idling. The size of the main jet, and the idling mixture strength and speed, can be adjusted to give smooth and economic running.

Diaphragm carburettors

Special carburettors have to be used for engines which do not always work at or near the horizontal. Engines for chain saws and some mowers are included in this category. The ordinary float chamber which overflows when used at an angle has to be replaced by a chamber of different design (Fig. 7.7).

As the engine piston moves on the induction stroke the air flow

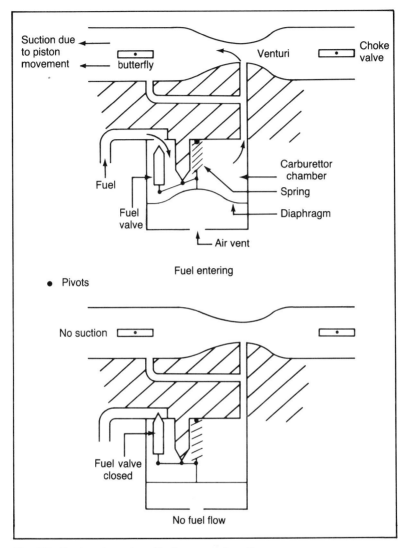

Fig. 7.7 Section through a diaphragm carburettor.

through the venturi sucks fuel from the carburettor chamber through the jet. Atmospheric pressure on the side of the diaphragm open to the air bulges it into the chamber and causes the fuel valve to open, allowing more fuel into the chamber. On the other engine strokes when there is no air flow and no suction the spring pushes out the diaphragm, closing the fuel valve and leaving the chamber full of fuel ready for the next induction stroke.

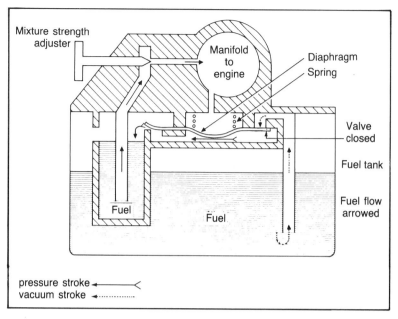

Fig. 7.8 Section through combined fuel tank and carburettor.

The complete fuel system

The fuel tank is usually fitted with an on/off tap, and fuel is supplied to the carburettor through a pipe which may include a filter. This may be no more than a copper or nylon gauze incorporated in the union where the supply pipe is connected to the carburettor. The fuel tank on most small horticultural engines is mounted higher than the carburettor so that the fuel can flow by gravity, but in some cases this may not be possible and a fuel pump may have to be used.

Figure 7.8 shows a common design of fuel system for small single-cylinder engines which combines the fuel tank and carburettor in one unit. It uses the pulsating air flow in the induction pipe to operate a simple diaphragm pump which transfers fuel from the main tank to a subsidiary internal chamber which acts as a float chamber.

Maintenance of the petrol fuel system

Maintenance of the fuel systems described is limited to cleaning the filters when fitted by washing them in fuel, keeping the hole in the fuel tank cap clear, and occasionally thoroughly cleaning the carburettor float chamber. If, despite these precautions, dirt should get into the carburettor the jets may become blocked. If so, they should be washed in fuel and blown out. A wire, or hand tool should never be forced through a jet in an attempt to clean it as this may enlarge the orifice.

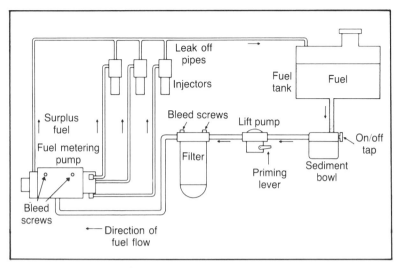

Fig. 7.9 Layout of the components of a typical diesel fuel system.

Fuel systems for diesel engines

In diesel fuel systems the carburettor is replaced by a fuel pump which accurately meters the exact amount of fuel required by each cylinder at any time and raises its pressure to the pressure necessary for injection into the cylinder. The metered fuel is distributed through steel pipes to an injector in each cylinder. The injectors open when injection pressure is reached and atomize the fuel into minute droplets. The interior of the cylinder head, and the piston crown, are designed to ensure thorough distribution of the droplets throughout the combustion space so that all the fuel may be completely burnt (Fig. 7.9).

The remainder of the fuel system is similar to that for a petrol engine, although more thorough filtration of the fuel is required as the precision finish of the pump and the injectors can be damaged even by minute dirt particles. Either one or two large filters with replaceable elements are fitted. In addition, a sediment bowl intended to trap water and coarse grit may be included in the system.

Maintenance of the diesel fuel system

Maintenance of the diesel fuel system is very important. It includes regular filter replacement and cleaning of the sediment bowl. After this service the system has to be bled to remove entrapped air. This maintenance should not be attempted without practical instruction. On no account should the pump or injectors be tampered with.

Governors

A governor is incorporated in almost every small engine and tractor engine used in horticulture. Its job is to maintain a steady engine speed irrespective of increases or decreases of load on the engine, and to free the operator from having constantly to adjust engine speed to counteract these changes in load. The foot or hand speed control that the operator moves to select the engine speed required is, in reality, altering the governor setting, and it is the governor which is linked to the carburettor butterfly or the injection pump which changes the engine speed.

Questions

1. Why is an air cleaner necessary on a horticultural engine?
2. What types of air cleaner are commonly used on small horticultural engines?
3. How is each type of air cleaner maintained?
4. What is the basic working principle of a carburettor?
5. What maintenance does a carburettor require?
6. What are the components of a fuel system for a diesel engine?

8 Ignition systems

The purpose of the ignition system of a petrol engine is to provide a spark in the engine cylinder to ignite the mixture of fuel and air at the correct point in the working cycle. The spark is produced by an electrical circuit which causes a spark to jump across the gap between the two electrodes of a sparking-plug. The plug is fitted into the cylinder head and protrudes into the cylinder. A very high electrical pressure of between 4 000 and 15 000 V is necessary to make the spark jump the gap between the electrodes.

In a car or truck ignition system this high voltage is produced by a coil-ignition circuit using a battery as its initial source of electrical pressure. The ignition system used for small engines has first to generate an electrical supply, and then the low voltage generated has to be stepped up to the high voltage required at the plug. This two-stage process is carried out by the magneto.

Magneto ignition

The electrical principles underlying the generation of electricity within a magneto is that whenever a coil of insulated wire is spun in the magnetic field which exists about a magnet a voltage is induced in the coil, and if this is connected to a circuit a current will flow. Similarly, if a magnet is moved rapidly close to a coil of insulated wire a voltage will again be induced in the coil (Fig. 8.1).

The flywheel magneto

In the simplest form of magneto used on small single-cylinder engines, magnets are fitted inside the rim of the flywheel and a stationary coil of copper wire wound on a laminated soft-iron core is fixed on the side of the engine casing. The ends of the coil are connected to a contact breaker which may be described simply as a switch operated by the flywheel as it rotates. The coil and contact-breaker circuit are referred to as the primary, or low-voltage, circuit. As the engine flywheel rotates, the magnetic field around the magnets passes the poles of the coil and induces a voltage in it, and a current flows through the primary circuit whenever the contact-breaker switch points are closed and the circuit is complete.

Other electrical principles have to be introduced to explain the step up in voltage from the low voltage induced in the coil to the high voltage required at the plug.

If a second coil of insulated wire is wrapped around the primary coil,

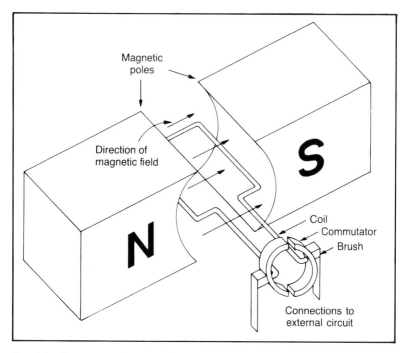

Fig. 8.1 Generation of current by spinning a coil in a magnetic field.

but is not in any way connected to it electrically, a voltage is induced in it every time the primary circuit is switched on or off. Furthermore, the voltage induced in the second coil will depend on the relative number of turns of wire in the two coils. If there are a thousand times as many turns of wire in the second coil as there are in the first, the voltage will be stepped up a thousand times. Thus it is possible to step up the voltage in the second coil to that required to produce a spark at the plug. The second coil is connected to the sparking-plug electrode, and this circuit is referred to as the secondary, or high-tension, circuit.

Thus a low voltage is induced in the primary coil as the flywheel rotates, and this is stepped up to the voltage necessary for ignition every time that the contact-breaker points separate.

The contact-breaker points are opened by a cam at the centre of the flywheel. This moves one of the points and causes them to separate quickly, and close again more slowly, once every revolution, and just at the right moment in the engine cycle. The stationary point is earthed to the engine to complete the circuit from the coil.

To simplify the circuit wiring, the metal of the engine is used as a conductor to complete each of the circuits, and so the only obvious connections are from one end of the primary coil to the contact breaker, and from one end of the secondary coil to the plug. The other end of each coil is connected to the metal of the engine. These connections are

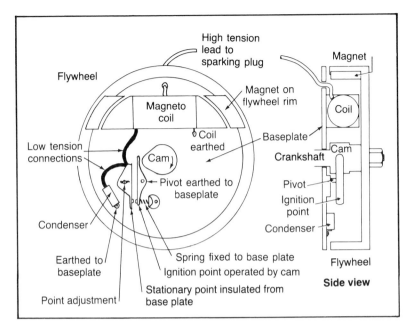

Fig. 8.2 Two sections through flywheel magneto.

said to be earthed, although this must not be confused with the earthing of an electrical mains installation which acts as a safety measure and is not part of the working circuit.

It will be noted from the diagram (Fig. 8.2) that a condenser is included in the primary circuit and that this too is earthed. This appears to offer an alternative route for the current when the contact-breaker points are open. In fact this is not so, and the current does not continue to flow when the points are open. Instead, the inclusion of the condenser in the circuit prevents a large spark jumping across the contact-breaker points each time they open which would cause pitting and deterioration of the points.

Independent magnetos

In some cases the magneto is a self-contained unit, separate from the flywheel, and driven through gears from the engine crankshaft. In these magnetos a magnet may again be rotated near the coil, but sometimes the positions are reversed and it is the coil which is turned between the poles of a horseshoe magnet. There is no difference in the current produced.

The sparking-plug

The sparking-plug, which is screwed into the cylinder head, has a central electrode surrounded by ceramic insulating material. This

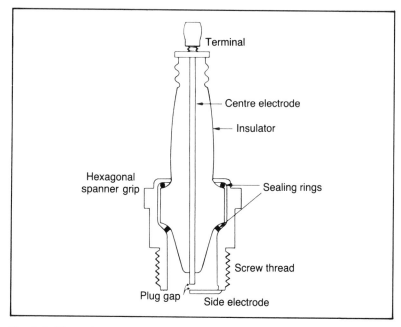

Fig. 8.3 Vertical section through sparking-plug.

is automatically earthed to the engine when the plug is screwed into the cylinder head. The high-tension lead from the secondary coil is connected to a terminal on the central electrode (Fig. 8.3).

Maintenance of the ignition system

Regular maintenance of the ignition system is essential. The frequency of attention depends on the engine type and duty but servicing is advisable at least every three months.

Maintenance of the sparking-plugs

The sparking-plugs must be removed and the electrodes cleaned. This can be done by sand-blasting, although over-zealous cleaning by this means will wear away the points and some manufacturers of plugs do not recommend it. When clean, the gap between the electrodes can be reset by bending the outside electrode, never the central one. A feeler gauge should be used to check the gap between the electrodes, which usually should be between 0.30 and 0.75 mm (0.015 to 0.030 in). When the electrodes are worn down, or if the ceramic insulation becomes cracked or chipped so that it no longer provides good insulation, the plug must be replaced. Care must be taken to ensure that the replacement plug is the correct type.

Maintenance of the contact-breaker points

Contact-breaker points must also be cleaned regularly. If they are pitted

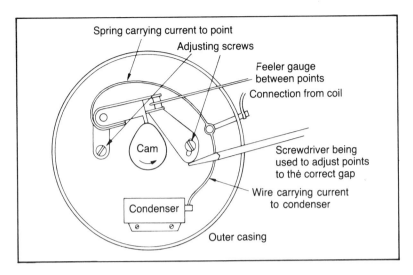

Fig. 8.4 Adjusting contact-breaker points.

or irregular they can be smoothed with a fine carborundum stone or ignition file. Care must be taken to ensure that the cleaning is not overdone and that the faces are kept parallel to each other. When clean, the gap between the points must be checked with the operating cam turned into the position where the points are at the maximum distance apart. The gap recommended is normally between 0.30 and 0.50 mm (0.012 to 0.020 in). The gap is adjusted by slackening off a screw which holds one of the points firm. This enables the point to be moved until the desired gap is obtained. The screw is then retightened, and the gap checked to see that the point has not slipped. Ultimately these points will have to be replaced when they can no longer be cleaned and the gap set correctly (Fig. 8.4).

The whole system must be kept clean and dry, and occasionally other parts such as the condenser may have to be replaced, but most faults can be traced to the plugs or points.

Solid-state ignition systems

These are increasingly being used in place of the conventional magneto ignition systems described. The basic principles for the production of a primary current still apply, and the high-tension and low-tension coils are retained to step up the voltage, but the cam and contact-breaker points are replaced by an electronic switch and trigger.

Elimination of the points reduces the maintenance and removes one frequent source of faults in the ignition system.

Other electrical circuits

Although the electrical equipment of small engines is normally confined to the ignition system, tractors and other vehicles may have additional circuits for lighting, indicators, horn, and other accessories, as well as for starting the engine. These circuits normally operate at 12 V and obey the same principles set out in Chapter 3 for mains installations.

The battery

The electrical pressure for these circuits is derived from the battery. This is really a storage battery in which chemical changes take place that allow it to store electrical energy for a period, and later give it up when it is required to operate the starter motor or an electrical accessory. A generator driven by the engine produces the electrical energy and charges the battery. In the battery are a number of plates immersed in diluted sulphuric acid, the electrolyte (Fig. 8.5).

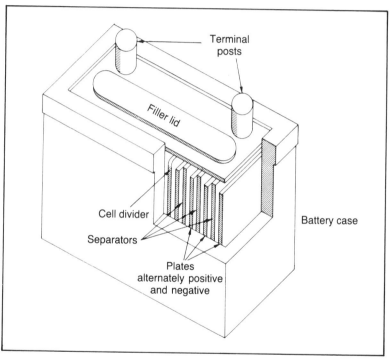

Fig. 8.5 Sectioned lead–acid battery.

One terminal post is marked with a positive (+) symbol, and the other with a negative (−) sign. On modern tractors it is usual for the negative post to be connected to the metal body structure which acts as a 'return' for all the different electrical circuits. When disconnecting the battery for any purpose the negative connection should be removed first, and then the positive. Reconnection should be in the reverse order.

Maintenance of the battery

The level of acid in the battery must never be allowed to drop below the top of the plates, and so distilled water must be added regularly to maintain the level. Any acid which splashes out of the top of the battery must be carefully wiped away. The vent holes in the filler plugs must be kept clear, and the terminals to which the wires are attached must be kept clean and coated with petroleum jelly. Connections must be tight, and the earth strap from the battery to the metal of the engine along with all other earth points must be checked to ensure that there is no corrosion causing resistance to the current flowing, a common fault of vehicle ignition systems. Low-maintenance batteries which require infrequent topping-up, or no topping-up at all, are being increasingly used.

Questions

1. How is the spark produced to ignite the mixture in a petrol engine?
2. What are the basic principles of a magneto?
3. Draw a diagram of a section through a sparking-plug.
4. What regular maintenance is required by a magneto-ignition system?
5. How are contact-breaker points gapped and set?
6. What is the purpose of a battery fitted to a tractor?
7. What maintenance is required by a battery?

9 Transmission

The drive from the engine of a tractor is taken to the wheels through a transmission system. This consists of a clutch, a gearbox, a differential, and usually separate speed-reduction gears. In addition, the drive normally has to be turned through a right angle to the back axle by means of further gears (Fig. 9.1).

The transmission of a pedestrian-operated machine may be very similar, although a differential may not be required, and some of the gearing may be replaced by belts or chains.

As an internal-combustion engine only develops its maximum power output over a narrow range of engine speed, and will not run at all below a certain idling speed, the transmission has to provide a range of forward speeds for any selected engine speed. The gearbox provides the range of forward speeds and reverse, perhaps as many as 12 forward speeds for a tractor, and 2 or 3 for a cultivator. The clutch allows the drive from the engine to be broken so that the machine can be stopped or a different gear ratio selected. It also allows loads to be taken up gradually. The differential allows the driving wheels to rotate at different speeds when the machine is turning a corner, but provides a positive drive to each wheel all the time.

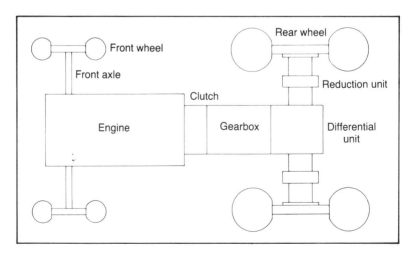

Fig. 9.1 Layout of tractor transmission.

The clutch

The principle of most transmission clutches is that the drive is transmitted by friction between two plates held together by pressure alone, one plate being driven by the engine, the other driving the remainder of the transmission. This principle is illustrated by Fig. 9.2 which shows the components of a simple clutch suitable for a tractor transmission.

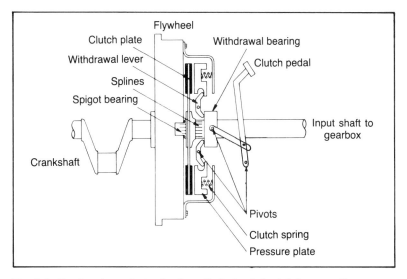

Fig. 9.2 Section through single-plate (dry) clutch (pedal depressed).

The thin, circular clutch plate is splined on to the shaft leading to the gearbox so that the clutch plate is free to slide along the shaft but cannot rotate without turning the shaft with it.

The clutch plate has linings or pads of special friction material made from asbestos fibre riveted or bonded to each side. This material can withstand the normal heat produced in a clutch without change in its properties.

When the clutch is engaged, the clutch plate is pressed against the flywheel by a spring-loaded pressure plate which always rotates with the flywheel. Thus the whole clutch assembly forms a sandwich, the two 'slices of bread' being driven by the engine, the 'filling' driving the transmission. When the clutch pedal is pushed down a linkage pulls back the pressure plate against the pressure of the clutch springs. This allows the clutch plate to edge along the splines away from the flywheel and turn freely, breaking the drive.

The same principle is used and the same parts are recognizable in clutches used on many other machines. Occasionally a number of plates are used which may be alternately of bronze and steel. Alternate plates are driving and being driven.

Clutch operation

Normally, cluches are operated by a foot pedal or hand lever, but centrifugal clutches are engaged automatically when a certain engine speed is reached. In these the plates are not held together by springs but by a linkage operated by weights which move centrifugally outwards as the engine speed increases and provides the necessary pressure to hold the plates together.

Dual clutches

In some models of tractor the simple single-plate clutch described is replaced by a 'dual' clutch. In addition to the usual transmission clutch, this incorporates a second clutch similar to the first which operates a 'live' power take-off. The two clutches are assembled one behind the other as a pack, and are usually controlled by a single clutch pedal so that the power-take-off shaft can be driven through one, and the wheels through the other. The first movement of the pedal disengages the drive to the wheels, but if the pedal is pushed right down the power-take-off drive is also disengaged. (Fig. 9.3).

The power take-off on many modern tractors is described as 'independent' and is operated by a separate hydraulically actuated clutch. The transmission clutch is the single dry-plate type illustrated in Fig. 9.2.

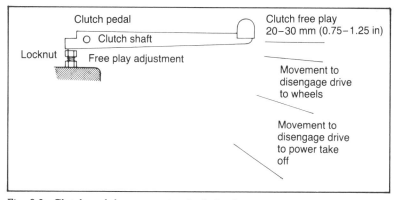

Fig. 9.3   Clutch pedal movement – dual clutch.

Clutch adjustments

Adjustments may have to be made to the clutch-operating linkage from time to time to ensure that there is some free movement before the clutch starts to disengage. Failure to make this adjustment to the manufacturer's recommendations may prevent the clutch from engaging fully and lead to clutch slip, the clutch overheating, and rapid wear of the lining materials on the clutch plate.

Dog clutches

Various forms of dog clutch are also used to engage and disengage

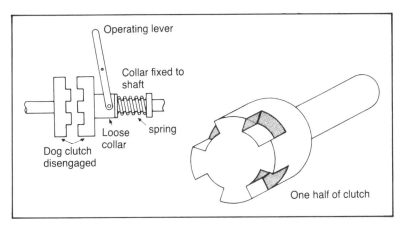

Fig. 9.4 The dog clutch.

drive, but normally this can only be done at rest and they cannot take up drive gradually. They too consist of two plates held together by springs, one driving and the other being driven. But the drive does not depend on friction as each plate has teeth which mesh together. Sometimes the teeth have sloping edges which slip over each other if the drive is overloaded so that the clutch doubles as an overload safety-release clutch (Fig. 9.4).

Gearboxes

The gearbox part of the transmission usually relies on gears, but drive may be by belts or chains. In each case the principle is the same. Whenever a small gear is used to drive a bigger one, or a small belt pulley or chain sprocket is used to drive a bigger one by a belt or chain, a reduction in speed results. The reduction which occurs depends on the relative number of teeth on the two gears or sprockets, or on the relative diameters of the belt pulleys. The relationship between the speed of the driving mechanism and that of the driven can be summed up as

$$\begin{matrix} \text{Speed of driven} \\ \text{gear or sprocket} \end{matrix} = \begin{matrix} \text{Speed of driving} \\ \text{gear or sprocket} \end{matrix} \times \frac{\begin{matrix} \text{Number of teeth on} \\ \text{driving gear or sprocket} \end{matrix}}{\begin{matrix} \text{Number of teeth on driven} \\ \text{gear or sprocket} \end{matrix}}$$

and similarly for belt drives:

$$\begin{matrix} \text{Speed of driven} \\ \text{pulley} \end{matrix} = \begin{matrix} \text{Speed of driving} \\ \text{pulley} \end{matrix} \times \frac{\text{Diameter of driving pulley}}{\text{Diameter of driven pulley}}$$

The gearbox provides different forward speeds by using different pairs of gears, belt pulleys, or occasionally chain sprockets, for each different forward speed required.

Figure 9.5 shows a three-forward-speed gearbox with the second gear engaged. Moving the gear lever on the gearbox will slide one of the engaged gears out of mesh and then slide the selected alternative

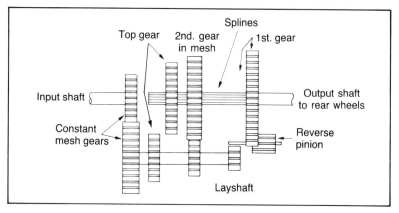

Fig. 9.5 Three-speed and reverse gearbox.

into mesh. One of each pair of gears is splined to its shaft to allow movement.

On many horticultural machines and some tractors, gears should only be selected when the transmission is stationary. Most tractors are fitted with a syncromesh gearbox which enables gear changes to be made on the move.

Reverse is obtained by introducing an extra gear between one of the pairs of gears which normally mesh together to give a forward ratio.

Gearboxes require no maintenance other than regular checks on oil level, and oil changes at the intervals recommended by the manufacturer.

Hydrostatic transmission

Hydrostatic transmission is an alternative to the conventional gearbox. It provides a stepless transmission which is operated by a foot rocker-pedal or hand lever which enables any forward or reverse speed to be selected without the use of clutch or gear levers. Most commonly it is fitted in smaller tractors, and it is very convenient when using a fore-end loader or other equipment which requires frequent adjustments in forward or reverse speed to be made. For instance, when operating a power-driven mower, forward speed can be adjusted – without stopping – to deal with a thick or tough area of grass without altering the engine speed.

The gears are replaced by a hydraulic pump driving a hydraulic motor. Moving the rocker-pedal alters the pump output and thus the motor speed giving a different ratio.

Belt and chain drives

Belt drives usually incorporate vee belts which have a wedge-shaped cross-section and run in vee-shaped grooves in the pulley rims. Drive is by friction between the sides of the belt and the sloping sides of the pulley grooves, never with the bottom of the groove (Fig. 9.6).

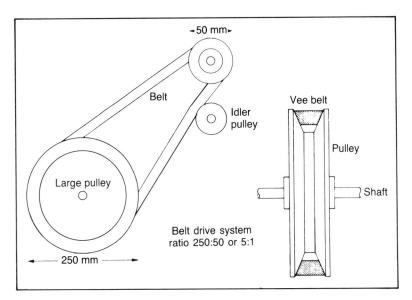

Fig. 9.6 Section through vee belt and pulley.

Apart from keeping the belts free from oil and out of strong sunlight, belt-drive maintenance is limited to keeping the belts correctly tensioned. This may be done by adjusting a free-running idler, or jockey, pulley. In very simple transmissions, tensioning the belt by means of a jockey pulley moved by a handle is sometimes used as an alternative to a clutch (see Fig. 15.22).

Chains should also be retensioned regularly, and where they run in an oil bath the oil level should be regularly checked and the oil changed at the intervals recommended by the manufacturer. Chains running in the open need light oiling or greasing. Occasionally, they need to be removed and immersed in a tin of paraffin for thorough cleaning before re-oiling and replacing. Chains working near the soil or in dusty conditions are normally left unlubricated, as a mixture of lubricant and soil can act as an abrasive and ruin the chain.

The tractor final drive

Where the drive has to be turned through an angle two special gears with their teeth cut at an angle to their faces are used. In a tractor transmission these gears are known as the crown wheel and pinion, and as the pinion is much smaller than the crown wheel they also provide a speed reduction.

The differential of a tractor is normally included in the gearbox housing, and requires no separate maintenance.

The normal action of the differential can be overridden by the differential 'lock' which locks both driving wheels together. In difficult working conditions when one wheel is tending to spin on an exceptionally wet patch, the use of the differential lock makes both wheels turn at the same speed, preventing all the drive from going to the wheel that is spinning and stopping the forward movement of the machine.

The differential lock is fitted to tractors and some pedestrian-operated rotary cultivators. On a tractor it is engaged by depressing and holding down a spring-loaded foot pedal. When it is engaged turning the tractor is almost impossible, but if the driver lifts his foot the pedal springs back up releasing the lock. Occasionally it may stay in engagement, but momentarily pushing down the clutch pedal or one independent rear brake pedal should free it.

Wheel equipment

Wheels

To provide maximum wheel grip and traction the driving wheels are large in diameter and are fitted with wide tyres. The bars or 'lugs' on the tyres which increase the traction are arranged in a vee pattern. Tyres should be fitted so that the point of each vee enters the soil first as the wheel turns. This forces the vee to close slightly. As the tread bars come out of the ground they flex outwards again and any soil trapped between them tends to drop out giving a self-cleaning effect.

Tyres on steering-wheels have continuous bands of rubber around their circumferences which prevent them from sliding sideways and makes steering more precise.

Studded tyres which do not mark the turf are available for use on golf-courses and sports fields as an alternative to the normal heavy-treaded tyres, and are fitted to both front and back wheels.

Wheel grip can be improved by adding weight to the rear wheels. This can be provided by bolting iron weights to the wheel centres or by filling the inner tubes of the tyre with a solution of calcium chloride and water. The calcium chloride is added to prevent freezing. A solution of 0.8 kg per 10 litres of water provides protection at temperatures down to -7 °C.

Weights may need to be added to the front of the tractor to prevent the front from rearing and to retain steering control when the tractor is carrying heavy rear-mounted implements.

Tyre pressures recommended depend on the type of work that the tractor or machine is doing, and the load that the tyre is carrying. A table which shows typical tractor manufacturers' recommendations is given below.

	Front		Rear	
	bar	lbf/in²	bar	lbf/in²
General use	1.4–2.0	20–28	0.82–0.96	12–14
Road work	1.8–2.1	26–30	0.96–1.1	14–16
Field work	1.4–2.0	20–28	0.69	10
On grass	1.4	20	0.69–0.82	10–12
With fore-end loader	3.1–4.0	45–60	0.82–0.96	12–14

However, increasingly, recommendations relating tyre pressures solely to load are being made by more tyre manufacturers.

High tyre pressures can lead to harmful compaction of cultivated soils and turf. Heavy tyre loadings and high tyre pressures should be avoided if possible, particularly if the ground is very wet. Tyre pressure gauges calibrated in bars are not always available, and so pressures in imperial units have been included in the table. Occasionally gauges marked in kg/cm² are to be found. This non-SI unit is approximately equal to 1 bar.

Tyres should be treated with care. Their life will be shortened if they are driven over sharp stones or projections, or are used under-inflated.

Tyre care

Weight should be taken off the tyres of stored machinery by supporting the machine securely on blocks, and the tyres should be shaded from the sunlight which deteriorates the rubber. Oil and grease should also be kept off tyres as these too have harmful effects on rubber.

Wheel track

The wheel track on tractors and cultivators can be adjusted to the width required to match different inter-row cultivation machines and operations. For instance, the distance between tractor wheels can frequently by adjusted in 100 mm steps from 1.2 to 1.9 m.

Brakes

The two rear wheels on a tractor are fitted with brakes which can be operated independently by separate pedals to enable sharp turns to be made when manoeuvring.

Brakes should not be used independently more than is necessary, and never in the higher gears. For road use, or when travelling at speed, the two pedals can be locked together so that both brakes are applied simultaneously. Brakes must be tested regularly, and particularly after repeated use of one independent brake, to ensure that both brakes go on at the same time when locked together. Testing is done by simultaneously applying both brakes sharply when driving steadily on an area of level ground. If the tractor slews in one direction, the brake on that side is adjusted too tightly, or the opposite one too slackly. Uneven braking can be rectified by adjusting the linkage of each brake in accordance with the manufacturer's instructions until the tractor can be braked without tending to slew in either direction. When correctly adjusted it should be possible to depress the brake pedals by approximately 18 mm (0.75 in) before the brakes are applied. This provides a running clearance to prevent the brakes continuously rubbing and overheating.

The brakes can usually be locked in the 'on' position for parking by a latch on the foot pedals, or by a separate hand lever. Occasionally a transmission brake locking the transmission is used for parking. Like the clutch, brakes rely for their operation on the principle that when two moving surfaces are pressed together there is frictional 'drag' between them.

Drum brakes

Drum brakes are fitted on some tractors and also on trailers. In its

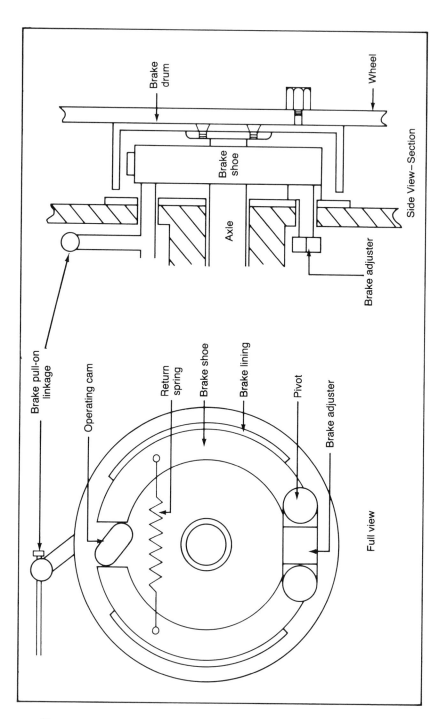

Brake drum

Wheel

Brake shoe

Axle

Brake adjuster

Side View–Section

Brake pull-on linkage

Operating cam

Return spring

Brake shoe

Brake lining

Pivot

Brake adjuster

Full view

simplest form a drum brake consists of two shoes with friction material bonded or riveted on them, each pivoted at one end to a circular plate fixed to the body of the tractor. When the driver pushes down a brake pedal the interconnecting linkage turns the brake lever, as shown in Fig. 9.7. This forces the shoes outwards against the inside of a circular brake drum which is fixed to the wheel. The friction causes the braking action.

Disc brakes

Disc brakes are also commonly fitted on modern tractors. These must not be confused with the disc brakes fitted on cars which are different. Tractor disc brakes are not fitted on the wheel but act on part of the axle leading from the differential to the reduction unit (see Fig. 9.1).

Two circular brake discs faced on each side with friction material are splined on the primary shaft and enclosed in a fixed housing which is part of the axle casing. When the brake is applied, a 'spreader' forces the two discs apart and presses them against the two end faces of the housing. Friction occurs at the points shown in Fig. 9.8 and causes the braking action. As a variation on this design the friction-faced discs are replaced by a series of metal plates immersed in oil.

Each type of brake has springs to pull back the shoes or discs when the brake is released.

Steering

The front wheels which steer the tractor usually turn on short stub axles which are linked by a track rod. Movements of the steering-wheel by the driver are changed in the steering-box into backward and forward movements of a drop arm. This movement is transmitted through the drag link to an arm which turns one of the front wheels (Fig. 9.9).

Alternatively, each front wheel can be turned by its own separate drag link and arms. These move in opposite directions, one backwards and one forwards, when the steering-wheel is turned. This system allows the track rod to be omitted and simplifies the alteration of the front-wheel track.

Normally, if the distance between the wheel rims at hub height in front and behind the axle are compared, it will be found that the wheels tend to point slightly inwards at the front. This is known as 'toe-in', and is adjustable. If the toe-in recommended by the manufacturer is not maintained, rapid tyre wear will occur, and so toe-in should be checked from time to time and, if necessary, adjustments should be made.

Power steering

Power steering is fitted to many modern tractors. The drag link and drop arm shown in Fig. 9.9 are replaced by a small double-acting hydraulic ram (a ram which can 'pull' or 'push' – see Ch. 10). Turning the steering-wheel to left or right meters oil under pressure to one end of the ram or the other, and this results in corresponding movements of the front wheels. The oil pressure is created by a pump driven by the engine, but if pressure should be lost for any reason turning the

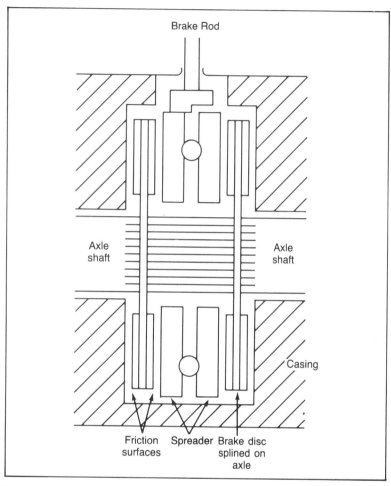

Fig. 9.8 Section through a disc brake.

steering-wheel will still turn the front wheels, although all the power assistance will be lost.

Questions

1. What are the main components of a transmission?
2. What is the purpose of each component?
3. What is the basic working principle underlying the single-plate clutch?
4. What is a dual clutch?
5. When are dog clutches fitted?
6. If a gear with 30 teeth turning at 100 rev/min (revolutions per

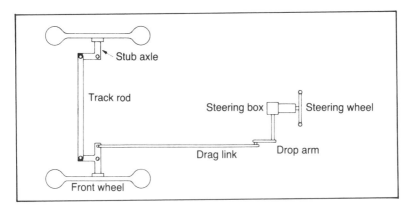

Fig. 9.9 Layout of steering gear.

minute) is driving a gear with 60 teeth, how fast will the second gear be rotating?

7. Draw a simple gearbox layout indicating which gear ratio is given by each pair of gears.
8. How are tyres self-cleaning?
9. What care should be given to tyres?
10. What is the procedure for testing independent brakes to see if they are working equally?
11. What is 'toe-in'?

10 Tractor hydraulic lift and power-take-off drive systems

Not only is the tractor intended for pulling implements but it also has been designed to be able to carry implements and loads and to be able to drive and operate machines attached to it.

Tractor hydraulic systems

The hydraulic lift system built into most tractors enables an implement to be quickly and easily attached to three links at the rear of the tractor. Holes in balls incorporated in the ends of the links fit over matching pins on the implement. The balls allow movement to occur between the tractor and the attached implement or machine. The matching holes and pins are made in several standard size categories to facilitate interchangeability of equipment and tractors. Category I is intended for small tractors and low-draught implements, whereas the bigger category II is for larger tractors and implements (Fig. 10.1).

Fig. 10.1 The rear of a tractor showing the hydraulic linkage.

Layout of the hydraulic system

A basic hydraulic system consists of a pump driven from the transmission of the tractor, or mounted on its engine, which pumps oil to a hydraulic ram. This consists of a cylinder with a close-fitting piston in it, rather like an engine cylinder. As the oil is pumped into the closed end of the cylinder the piston is forced along it. The movement of the piston is transmitted to the lower links by means of a cross-shaft and lift rods. A control valve controls the flow of oil, and can direct it back to the reservoir and allow the oil in the ram to flow out again when the links are to be lowered. The valve can also trap the oil in the cylinder when the links are to be held at any height (Fig. 10.2).

Most tractor hydraulic systems work at a maximum oil pressure of between 140 and 190 bar. This is limited by the pressure control valve.

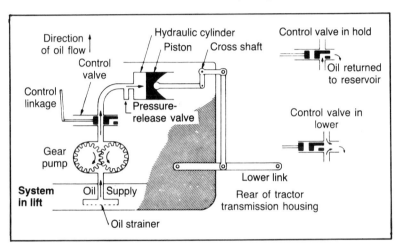

Fig. 10.2   Section through hydraulic-lift system of a tractor.

Maintenance of the hydraulic system

The transmission oil is frequently used for the hydraulics, and a filter is often incorporated into the system to clean the oil and to remove dirt and irons filings from it. Apart from servicing the filter at the intervals recommended by the manufacturer, the hydraulic system requires no maintenance.

Hydraulic control systems

Position control

The control valve can be operated directly by the driver to raise or lower an implement mounted on the linkage, and to hold it at any chosen height. This system is known as position control.

Draught control

A second, more complicated control system known as draught control is also fitted in most modern tractors. This system enables the working depth of an implement to be controlled continuously without the need for a depth wheel on the implement. The hydraulic control valve reacts to changes in the compression force in the top link, or tension in the lower links, which are due to changes in the draught (pull) required by the implement. Using the draught-control operating level the driver can preset the hydraulics to allow the implement to work at a certain depth, which also results in a certain draught. If the implement runs at too shallow a depth, the reduction in draught is sensed through either the top link or the lower links, and the hydraulic system control valve allows the implement to sink to its preset depth. If an implement goes too deep the draught increases and this is again sensed through the top link or lower links. The control system then raises the implement until the draught is back to the preset level and the implement is at the required depth again (Fig. 10.3)

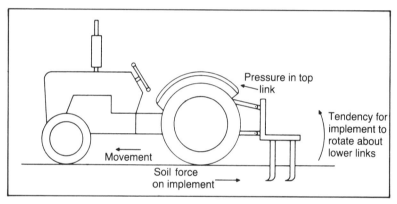

Fig. 10.3 Draught-control system.

Changes in soil type in an area being cultivated will result in changes in draught unrelated to changes in depth. The driver may have to modify the original depth setting to counteract these and to maintain an even depth of work.

Response control
The response control is an additional control on the hydraulic system which enables the rate at which the lower links lower and/or lift to be varied.

Attaching an implement
To attach an implement, the tractor is reversed squarely up to the implement and the lower links are raised or lowered by position control until they are at the right height for the left lower link to be fitted over the corresponding pin on the implement and secured with a linchpin. Next, the right lower link is attached and secured in a similar way, although small changes in its vertical position are made by using a crank handle on the lift rod, and not by using the control levers. The top link is adjustable in length and can be adjusted until it too can be

attached and secured. Adjustable check chains fitted to the lower links limit the sideways movement of the linkage and prevent the links and implement fouling the rear wheels. A stabilizer bar can be fitted between one lower link and the axle casing to limit any sideways movement of mounted equipment, particularly when accurate positioning is required.

External hydraulics

Tapping-off points connected to the hydraulic system are provided at the rear of the tractor. Rams, and even small hydraulic motors, fitted to machines trailed behind the tractor or mounted on it can be operated by oil fed through a flexible connecting pipe, or pipes. Hydraulic motors rotate when supplied with a continuous flow of oil under pressure from the hydraulic pump of the tractor, but require a second connecting pipe to return the oil from the motor to the tractor.

Power-take-off drives

Engine power can be used to drive machines mounted or trailed by the tractor through a power-take-off shaft which normally emerges from the rear of the tractor, although additional take-off positions are provided on some tractors. The shaft is driven from the transmission as explained in Chapter 9. To comply with the relevant British Standards a power take-off must turn at 540 ±10 revolutions per minute. The engine speed at which the standard power-take-off speed is given is usually marked on the proofmeter mounted on the instrument panel of the tractor (Fig. 10.4).

A two-speed power take-off is fitted to many tractors, and a three-

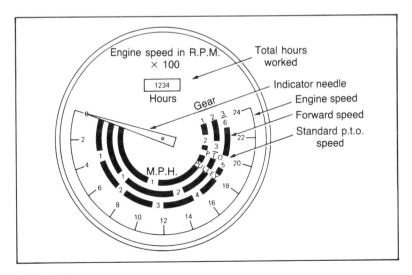

Fig. 10.4   Proofmeter.

speed to a number of small models. The second speed provided commonly gives 1 000 rev/min.

In order to allow movement between tractor and machine, the detachable power-take-off shaft which continues the drive from the tractor to the machine articulates at two points through universal joints, and is telescopic. If the two halves of the telescopic shaft are not pushed together with the universal joints in the same plane, as shown in Fig. 10.5, serious vibration and damage may occur.

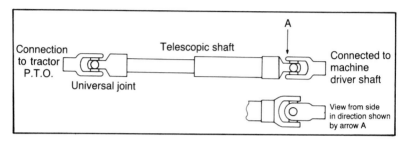

Fig. 10.5   A power-take-off shaft without guards.

Statutory safety regulations demand that at all times when a power take-off is connected the telescopic shaft and the universal joints must be completely guarded, and the guards must not be removed by anyone under the age of sixteen. New guards will have to be anchored so that they cannot rotate. Even when the power-take-off shaft has been disconnected the short splined stub protruding from the tractor has still to be guarded by a cover or a shield capable of withstanding a load of 113.4 kg.

The power-take-off must be disengaged and the tractor engine stopped before any adjustments are made to a machine.

Maintenance

Usually no maintenance of a power-take-off shaft is necessary other than regular checks to ensure that the safety guards are undamaged. Some shafts are fitted with grease nipples on the universal joints and on the telescopic section which require attention each time the shaft is used. Worn joints require replacement.

Questions

1. What is the correct procedure for attaching a mounted implement?
2. Draw the layout of the components in a tractor hydraulic-lift system.
3. What is draught control?
4. What is position control?
5. What is the British Standard Speed which applies to the turning speed of a power take-off?
6. What maintenance is required by power-take-off shafts?

11 Horticultural hand tools

Hand tools are widely used in horticulture because many of the areas cultivated are too small to allow efficient use of machines. Correct selection and use of these tools is important if good work is to be done (Fig. 11.1).

The spade

The spade is used for digging, trenching, and removing soil. It should not be used as a lever for large stones as this may strain the handle and damage it. Spades are available in two common sizes. The larger spade, usually known as a digging spade, has a blade 280 mm long and 190 mm wide and weighs between 2 and 2.5 kg. A smaller border spade has a blade 220 m long and 140 m wide and weighs between 1.5 and 2 kg. It is most important that the correct size is used, as a spade which is too large will make the work hard. Spade handles are made out of ash which is strong but light and the end may be 'D' or 'T' shaped. The 'D'-shaped handle is more comfortable to use. Some modern spades have plastic grips and the shaft is covered with a plastic skin to protect it from the weather. The blade is made from tempered steel and the top edge, or 'tread' should be wide so that it does not cut the worker's boot. Stainless steel blades are used on some spades. This makes them more expensive but easier to work.

When digging, the blade should be kept vertical so that it will cut through the soil with the minimum of effort.

The fork

Forks vary in type and use.

Digging forks are used for digging over the soil already turned by the spade. They normally have four prongs which can be either round or square in section. Like the spade, forks can vary in size and weight (Fig. 11.1).

Border forks are intended for weeding in borders, and two forks can be used back-to-back for root division. They are narrower and lighter than digging forks and usually have three or four prongs.

Potato forks are used for lifting potatoes because the potatoes stay on the prongs while the soil drops through and so less damage is done to the tuber. Compost and rubbish can easily be cleared with a potato fork. The prongs are much broader than on the digging fork and are usually four or five in number.

Fig. 11.1   A selection of hand tools.
From left to right – a digging fork, a spade, a draw-hoe, and a rake.

Some forks have a tubular steel shaft covered with a plastic skin to make them lighter to use.

Rakes

Rakes, like forks and spades, vary in their size and shape (Fig. 11.1).

Garden rakes usually have 10 rigid teeth and are used after the fork to break up the soil, level it, and break it to a fine tilth. They can also be used for gathering stones, weeds, and hedge clippings. The rake should not be used for moving large quantities of soil but should be drawn lightly across the soil, care being taken not to dig the teeth in too deeply.

Larger aluminium rakes are now available which are light and are especially useful when raking larger areas.

Lawn rakes are used on lawns to rake out the dead grass and collect leaves, as its springing action is less severe on the grass. The rake consists of 20 or more spring teeth arranged in an arc.

Hay rakes are used for raking together grass and leaves. The rake is made out of wood and has about 12 pegs for teeth.

When not in use, rakes should never be laid down but stood up against a wall or fence so that the teeth point inwards. If laid down with the teeth upwards there is a danger of someone stepping on them.

Hoes

Hoes are mainly used for weed control and loosening the soil around plants.

The draw hoe (Fig. 11.1) chops through the roots of the weeds, especially tap-roots and can also be used for making drills for seed sowing. It is used with a pulling or pushing action and the worker, who stands beside the drill being hoed, may move either backwards or forwards.

The push or Dutch hoe is used with a pushing action as the operator walks backwards. The angle between the blade and the soil should be between 20 and 30 degrees for the best results to be obtained.

The triangular-headed hoe is used with a pulling action for making a shallow drill, usually in conjunction with a garden line.

Trowels

Trowels are small scoop-shaped tools used for making small holes for plants. Their curved blades, usually 125 to 200 mm long, cut the soil round the plant leaving a ball of soil around the roots. Cheap trowels have steel handles which tend to buckle easily, while better-quality trowels have wooden handles.

Garden lines

The garden line is a very important horticultural tool used for setting out straight lines. It is used when seed sowing, trenching, lawn edging, transplanting, and for many other garden operations. It consists of between 6 and 20 m of twine with a pin at one end and a reel at the other on which the twine can be wound for storage. Polythene or nylon twine is best as it is rot-proof and does not shrink if it gets wet. The pin and reel may be made out of wood or steel.

Shears

Shears are commonly of two types, hand shears and edging shears.

Hand shears are used for cutting grass and hedges. They have blades 200 to 300 mm long and wooden or plastic handles. The cutting action takes place between the two blades which are pivoted and the material to be cut is sheared between the two blades. The blade edges are ground at an angle just less than 90 degrees. When sharpening shears, it is very important that the original angle of the blades is maintained. Sharpening is usually done with a carborundum stone. Good shears have a pruning notch near the pivot of the blades, so that thick twigs can easily be cut.

Edging shears are used for cutting lawn edges, and because of the length of the handles they can be used when standing upright. They are similar to hand shears, but have handles about 1 m in length attached at an angle to the blades.

Secateurs

Secateurs are small hand cutters used for pruning bushes and shrubs. There are two common types, parrot-bill secateurs and fixed-blade secateurs.

Parrot-bill secateurs have two blades with curved edges. They pivot and overlap as cutting takes place between them. Sharpening the parrot-bill type is very difficult and needs extreme care.

Fig. 11.2   A turfing iron and turf lifter.

Fixed-blade secateurs have one sharp blade which cuts against a soft copper plate or anvil. The blade is much sharper than those of the parrot-bill type and is usually replaceable. In time a groove wears in the copper plate and this also requires renewing if a 'clean' cut is to be maintained.

Turfing iron

The turfing iron consists of a half-moon blade attached to a handle. It is used for edging lawns or cutting turf prior to lifting (Fig. 11.2).

Turf lifters

Turf lifters have a heart-shaped blade attached at an angle to a shaft. The angle is so designed so that the blade can easily be slipped under the turf so that a horizontal cut can be made (Fig. 11.2).

Care of tools

Tools used solely for their correct purpose and properly maintained will provide good service for many years.

1. All tools must be thoroughly cleaned after use and must not be put away wet.
2. Metal parts should be wiped with an oily rag to prevent rusting.
3. The edges of spades, hoes, turfing irons, and turf lifters should be kept sharp with a file.
4. Tools should be stored in racks and not thrown together in a heap. Storage in racks prevents the tools from getting broken and saves time when getting out a particular tool.
5. Handles and shafts should be inspected from time to time to see that they are not cracked. Broken or cracked shafts can be dangerous and cause splinters. They should be repaired or replaced.
6. Frequent sharpening of cutting tools will make them easier to use and should result in better work.

Questions

1. Name three types of fork.
2. Name three types of rake.
3. Why must tools be cleaned before they are put away?
4. How can the edges on spades and hoes be kept sharp?
5. Why must broken shafts be replaced?

12 Ploughs and cultivating equipment

Before crops can be grown the soil must be cultivated. Cultivations are necessary to break up the soil into small particles, known as a tilth, which provide the correct conditions for seed germination and plant growth.

One of the oldest primary cultivation techniques is that of ploughing. The original plough was just a piece of pointed wood which broke up the top few inches of the soil. Today the modern plough can cultivate to a depth of over 300 mm. In recent years there have been many alternatives to ploughing suggested, but the use of the plough still has some advantages.

1. It is the only implement that completely inverts the soil, bringing up fresh soil to the surface.
2. It buries surface trash and perennial weeds and kills annual weeds.
3. It exposes a larger surface area for frost action.

The plough

The modern plough is usually mounted on the hydraulic-lift system of the tractor, but a number of trailed ploughs are still available. The hydraulic-lift system enables the plough to be raised and lowered so making it easy to manoeuvre it into field corners and on headlands.

The function of the plough is to cut out and then invert a rectangular slice of soil, referred to as the furrow slice. It should trim off any weed on the corner of the slice so that when a second furrow is laid against its neighbour no green material is visible between the slices.

The slice to be turned is separated by a vertical cut made by a coulter and is partially undercut by the share. A small piece is left uncut to act as a hinge. The mouldboard which follows raises and inverts the slice. If the hinge is not left the slice will merely move to one side and not be inverted. (Fig 12.1).

All the working parts are mounted on a frame made of cast or tubular sections bolted or welded together. The plough bodies are attached to legs which project downwards from the frame, and the other components are attached directly to the frame. Figure 12.2 shows the basic parts of the plough, but on many modern ploughs some of the parts may not be present.

The coulter

The coulter which makes a vertical cut is either a knife or a steel disc which is free to rotate.

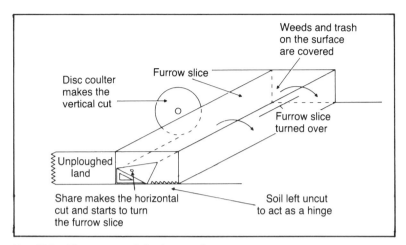

Fig. 12.1   The turning of the furrow slice.

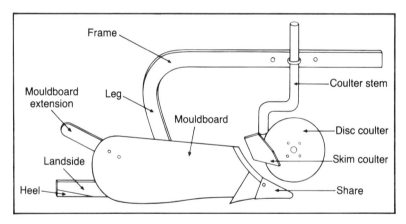

Fig. 12.2   The parts of the plough.

A cast-iron or cast-steel frog is attached to the bottom of the leg and is used as a mounting for all the soil working parts which make up the body.

The share

The horizontal cut of the furrow slice is made by the share which is triangular in shape and acts like a wedge, forcing its way through the soil lifting as well as cutting. The shares used to be made of chilled cast iron because of its cheapness but most are now made from cast steel, hardened to give extra wearing properties. Steel shares must be used where stones might fracture cast-iron ones, and are hard surfaced to improve their wearing properties.

The share is designed to penetrate the soil by what is known as

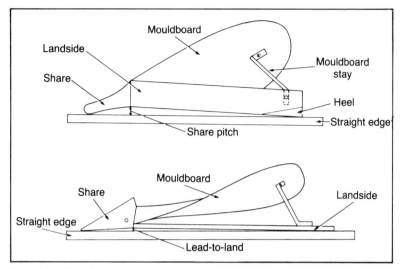

Fig. 12.3   The share pitch and lead-to-land of a plough body.

'share pitch' and to pull into the unploughed land by 'lead-to-land' (Fig. 12.3).

The mouldboard

This continues the lifting of the furrow slice started by the share and turns the soil over. It is usually concave in shape and its length depends on the type of body. Mouldboards are often made out of soft-centred steel. Fastened to the rear of the mouldboard is the 'mouldboard extension' which controls the amount that the furrow slice is turned over. Many modern types of mouldboards are now available to increase the speed of ploughing and reduce the pull required.

The landside

Landsides are usually made of steel. They press against the furrow walls and take the side thrust of the plough. The rear landside is usually longer than the others and has a wearing part at its end called the 'heel'.

Skims

Skims can be fitted to the disc coulters so that surface trash can be completely buried and grass and weeds do not grow through the furrow slices. The skim is set to pare off a corner of the furrow slice and direct it into the bottom of the furrow. The mouldboard then turns the furrow slice on top.

Body types

There are three types of body in common use (Fig. 12.4):

1. The general-purpose body is the longest of the three bodies and has

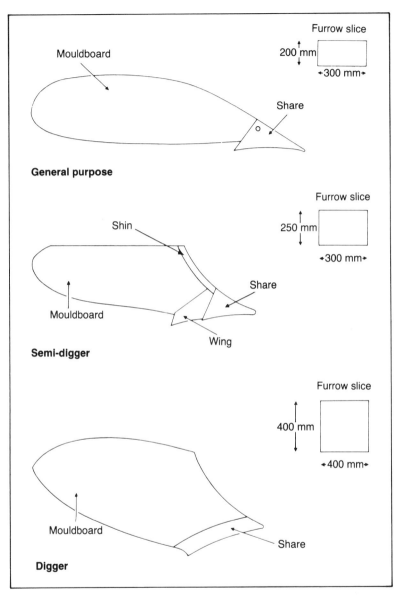

Fig. 12.4   Plough body types.

a gentle turning action giving an unbroken furrow slice which can be left over winter for the frost to weather. It can plough to a depth of 200 mm.

2. The semi-digger body is shorter with a deeper concave surface. It turns the furrow slice more sharply so leaving it broken, but speed

of operation can vary its action. Reducing the speed will produce a less broken furrow. The body is made up of four parts, the share, the wing, the shin, which is the replaceable leading edge of the mould-board, and the mouldboard. It can plough to a depth of 250 to 300 mm. The tilth left by this type of body presents the soil for weath-ering and also reduces the amount of after-cultivations necessary.
3. The digger body is shorter than the semi-digger body and can plough to a depth of 400 to 450 mm. It gives a very broken furrow slice and is widely used for many horticultural cultivations, particu-larly when a quick tilth is required.

Before ploughing the tractor wheel widths must be set to match the furrow width of the plough, the correct settings being given in the plough handbook.

Coulter settings

The disc coulter is set so that the hub of the coulter is above the share point. The distance between the edge of the coulter and the share for normal ploughing is 40 mm (two fingers). Between the side of the disc and the share the distance should be 20 mm (one finger) (Fig. 12.5).

The skim is set 50 to 75 mm deep, the tip of the skim as close to the coulter as possible, and the heel 5 to 15 mm away.

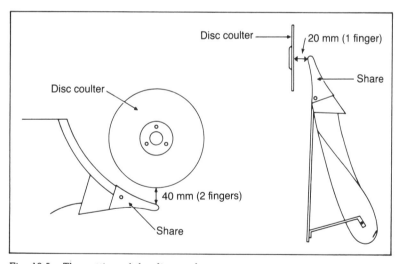

Fig. 12.5   The setting of the disc coulter.

Plough adjustments

1. *Depth.* The depth of ploughing can be adjusted in one of two ways:
    (a) By a depth wheel fastened to the plough. This can be raised or lowered by a handle connected to the cranked wheel axle and within reach from the tractor seat.
    (b) By the tractor hydraulics. Draught control enables the depth to

be adjusted and maintained constant on undulating surfaces without the need for a depth wheel. A fully mounted plough on draught control transfers the weight of the plough to the rear wheels of the tractor, reducing wheel spin and giving better traction.

2. *Front furrow width.* The width of the front furrow is very important if even ploughing is to be obtained. The front furrow width can be increased or decreased by 50 to 75 mm by rotating the cranked cross-shaft, which twists the plough in relation to the tractor.

3. *Front furrow depth.* This is altered by tilting the plough, using the adjustable right-hand-side lift rod on the hydraulic-lift system of the tractor. Lengthening the lift rod lowers the front furrow and vice versa. Too deep a front furrow can cause uneven ploughing.

4. *Pitch.* This is adjusted by lengthening or shortening the top link of the tractor. The length is correct when the heel of the rear body is just marking the furrow bottom. In some cases when penetration is poor it may be improved by shortening to increase the pitch.

Plough maintenance

1. All the moving parts, disc coulter bearings and adjustment levers should be greased daily.
2. All the components should be tight, especially the soil working parts.
3. Worn shares must be replaced if good ploughing is to be obtained.
4. After use the bodies should be painted with waste oil or preservative to prevent rusting so that the furrow slice slides over the mouldboard next time the plough is used.

Reversible ploughs

Reversible or one-way ploughs are now very popular. Unlike the conventional right-hand plough the reversible plough has two sets of bodies, one right-hand and the other left-hand. This enables the plough to turn furrows to left and right depending on which set of bodies are used. A turn-over mechanism swings the bodies over when the plough is lifted from work, and can either be manually or hydraulically operated. Mounted reversible ploughs are usually restricted to 3 furrows due to the extra weight of the additional bodies, and it is most important to add front-end weights to the tractor to enable positive steering to be maintained. The advantages of this type of plough are:

1. a saving in time;
2. the level finish produced;
3. their suitability for small areas of land.

Rotary cultivators

Rotary cultivators are widely used in horticulture as an alternative to ploughing. They are capable to producing a seed-bed in fewer operations than ploughing followed by normal cultivations. A 'fluffy' tilth is produced which should be allowed to settle before seeds are sown. Rotary cultivators can be either self-propelled and pedestrian-operated or tractor-mounted and power-take-off-driven.

Fig. 12.6 A pedestrian-operated rotor-driven cultivator.

Pedestrian-operated machines

There are many small pedestrian-operated machines on the market having a four-stroke engine as their source of power. They can either be self-propelled, using rubber-tyred drive wheels, or use the revolving rotors to pull the machine through the soil. The larger machines have a gearbox which enables the operator to select 2 or 3 forward speeds, and a reverse to help in manoeuvring. The speed of some small machines can be adjusted by altering a vee-belt drive. The drive from the engine is usually by vee-belt or chain and then by gears to the land wheels while a chain enclosed in a case takes the drive to the rotor. On machines not fitted with drive wheels, the type of tilth produced depends on the operator holding back on the machine or pressing down on to the rear foot. On hard ground quite a considerable amount of pull may be required otherwise the rotors will pull the machine forward too quickly and not produce a proper tilth (Fig. 12.6). It is an advantage to fit transport wheels on machines not fitted with drive wheels to enable easy transportation from one site to another.

The rotor is a shaft on which are bolted a series of L-shaped blades, which can be changed when worn. Around the rotor is a shield which controls the type of tilth produced and protects the operator from flying stones and clods. The blades rotate in the direction of forward travel and cut out slices which are pulverized against the shield and by other blades (Fig. 12.7).

The width of the rotor varies from machine to machine but is usually 300 to 1 200 mm. This is the working width of the machine. There are various types of rotor available and many machines have the facility to change from one type of rotor to another. Common types of rotor are:

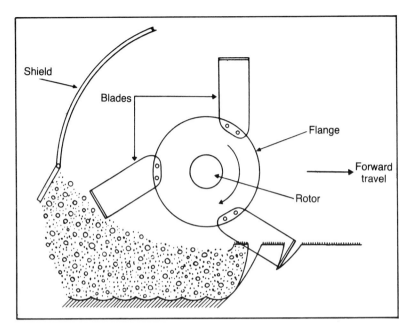

Fig. 12.7 The action of a rotary cultivator.

1. L-shaped blade rotor for normal cultivations;
2. finger rotors for seed-bed preparation;
3. slasher rotors for heavier work where a coarse type of tilth is required;
4. pick-rotors for breaking up soil which has 'panned' or hard-packed soil

Most of the controls are situated on the handles and include two clutches, one to break the drive to the land wheels, and another to engage and disengage the rotor. Some models have handles which can be moved to the side so that the operator need not walk on the rotovated ground.

Maximum depth of work is usually 150 to 200 mm and is controlled by an adjustable skid which enters the soil and prevents the rotor going too deep.

Tractor-mounted rotary cultivators

Larger tractor-mounted rotary cultivators driven from the power take-off are available where larger areas of ground are to be cultivated. Their operation is similar to that of the pedestrian-operated machines, but their working widths are considerably greater. Depth of working is usually controlled by a land wheel mounted on the side of the machine. The tractor gearbox is used to give a range of forward speeds because the engine speed must provide a standard power-take-off speed to the machine if it is to work correctly.

Adjustments for different tilth

The type of tilth produced depends on:

1. *the condition of the soil.* If the soil is too wet the rotary cultivators will not break up the soil but cut it into lumps;
2. *forward speed of the rotary cultivators.* Slow speed gives more cuts per metre;
3. *the position of the shield.* A raised shield will give a coarse tilth, a lowered shield a fine tilth;
4. *rotor speed.* On most small machines this depends on engine speed. The rotor speed of larger machines can be altered by interchanging gears in the drive, a high rotor speed gives a fine tilth.

Thus for a fine tilth a combination of a slow forward speed, which will give more cuts per metre, and a high rotor speed together with a low shield position is required. For a coarse tilth a combination of a faster forward speed and a lower rotor speed with a high shield position is required.

The rotary cultivator can be used for work other than seed-bed preparation.

1. If it is fitted with a row-crop rotor, the flanges on which the blades are mounted can be set for inter-row cultivations (Fig. 12.8).
2. It can be used for chopping up surface trash in reclamation work.
3. It can be used to control weeds between plots of ground or around the headlands.
4. It may be used for sports ground site construction or landscaping.

Some makes of rotary cultivator can have a range of attachments fitted to them so that they can be used for other jobs. These can include ridger bodies, a plough body for ploughing small areas, and a range of cultivator tines. Small trailers can be attached to some machines.

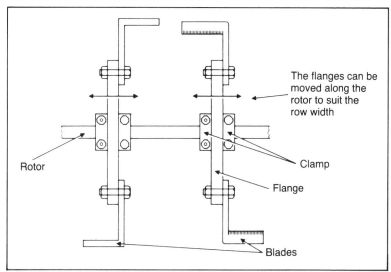

The flanges can be moved along the rotor to suit the row width

Rotor

Clamp

Flange

Blades

Fig. 12.8   A row-crop rotor.

Maintenance of the rotary cultivator

1. Maintenance of the engine should be carried out as described in Chapters 6, 7, and 8.
2. A daily check should be made to see that the bolts which hold the blades to the shaft are tight.
3. All moving parts should be oiled or greased before use.
4. The oil level in the chain case or gearbox should be checked and topped up with oil if necessary.
5. Any wire or trailing weeds should be removed from the rotor.

Spaders

Like rotary cultivators, spaders can cultivate to a full working depth and in some circumstances produce a tilth in one pass. They can bury and mix surface growth and manure to the full working depth.

The simplest form of spader is very like a rotary cultivator in principle except that the rotor turns more slowly and the blades are replaced by a smaller number of spade-like blades fitted on curved shanks. These give the digging action. There is no adjustable rear shield. The additional gearing necessary to give the secondary turning action complicates the mechanism considerably.

Tined cultivators

Cultivators follow the plough and are used to break up the furrow slice. They consist of tines which are drawn through the soil; the larger clods of soil are shattered on contact while some of the smaller clods are pushed aside. The tine points are tilted forward to pull them into the ground, so lifting some of the clods to the surface. The cultivator has a shattering and stirring effect on the soil. Care must be taken when using cultivators not to bring up buried trash or cold, wet soil.

Cultivators can be divided into three types:

1. rigid-tine cultivators;
2. spring-loaded tine cultivators;
3. spring tine cultivators.

The rigid-tine cultivator

This consits of a frame into which are fastened a number of vertical tines. Different types of points can be fitted to the end of the tines and these have different effects on the soil. The most common is the straight point which is usually reversible to extend its life and is used for general cultivations to a depth of 100 to 600 mm (Fig. 12.9).

The effect of the cultivator on the soil depends on the forward speed of the tractor, the number of tines, the type of point, and the condition of the soil. Weights may be added to the cultivator to improve penetration.

The spring-loaded tine cultivator

Used in stony soils which might damage rigid tines, the tines on this cultivator are held in position by strong springs and if they meet an

Fig. 12.9 A rigid-tine cultivator in action.
By permission of Ransomes, Sims of Jefferies Ltd.

Fig. 12.10 A spring-loaded tine cultivator in action.
By permission of Ransomes Sims & Jefferies Ltd.

obstruction they can spring back out of the way. When the obstruction
is cleared the springs return the tines to their original position. Their use
is similar to the rigid-tine cultivator but depth is limited to 225 to 300 mm
(Fig. 12.10).

The spring-tine cultivator

This has tines made out of spring steel. As the tine is drawn through
the soil it vibrates, shattering the clods better than rigid tines. The

number of tines varies from machine to machine. Some have a greater number of small spring tines and are used for shallow work. After two passes with one of these machines a rough seed-bed is obtained. Other machines have a few large spring tines which operate at a greater depth and in some cases can be moved along the frame for inter-row cultivations in potatoes (Fig. 12.11).

Fig. 12.11   A spring-tine cultivator in action.
By permission of the Ford Motor Co. Ltd.

It is common to cultivate in the same direction as the furrows for the first time and across the furrows for the second. The only maintenance on these types of machine is to check that both tines and points are secure before use. When the points are worn they need to be replaced. Many points are reversible and so provide two wearing parts.

Powered cultivators

Powered cultivators are often used to break down the soil to produce a tilth after ploughing. One type consists of a series of arms on which straight tines are fitted. The arms rotate in a horizontal plain causing the tines to pass through the soil in a circular motion. Power to the arms is from the tractor power take-off through a bevel gearbox. The effect on the soil depends on the speed of the tines in relation to the forward speed of the tractor. A slow tractor forward speed will give a fine tilth while a faster tractor forward speed will give a coarse tilth. A powered cultivator is capable of producing a tilth quickly and can remove the need to make several passes with other types of cultivation equipment.

Harrows

Harrows normally follow the cultivator in sequence for seed-bed preparation. They consist of small tines which break down the soil even more on impact and also cause consolidation of the lower layers. Their working depth is about 150 mm and they may consist of either rigid or spring tines. The tines are mounted in 4 or 5 rows on a frame and the number of frames pulled depends on the size of the tractor. The frame has a levelling action as it is drawn over the soil. There are usually about twenty tines on a frame. The frames are fastened to a pole which is pulled by the tractor. A small tractor can usually pull three or four frames.

Mounted harrows are quite popular because they are much more manœuvrable than trailed harrows.

Zigzag harrows

Varying according to their weight and length of tine, these are referred to as heavy, medium and light harrows, Heavy harrows are used to break down the soil after the cultivator while medium or light harrows can be used prior to sowing.

Spring-tine harrows

These produce a tilth by the vibrating action of their tines, the action being similar to that of the spring-tine cultivator. Depth is controlled by two land wheels, one on each side of the frame.

Chain harrows

These harrows consist of a mat of interconnected chain links. They have no rigid frame so that they follow the contours of the land. Some have a long and short tine on opposite sides to enable deep or shallow penetration to be obtained. This is determined by which way up the chain harrow is used. They are used for obtaining very level seed-beds and turf renovation.

Disc harrows (Fig. 12.12)

Disc harrows can be used as an alternative to cultivators when there is a great deal of surface trash. They consist of a series, or gang, of concave discs mounted on a shaft; four gangs usually being mounted on a frame as a 'set'. The discs are positioned equally along each shaft by spacers and an adjusting mechanism enables each shaft to be angled in relation to the direction of travel. As the disc rotates it cuts into the soil. The greater the angle on the disc the greater the penetration and the greater the throwing action of the disc. The discs throw a small furrow, and so in order to keep the land level the rear discs return the soil thrown out by the front discs. Penetration can be increased by putting weights on to the framework.

Discs may be either mounted or trailed. Sets, consisting of two gangs, and offset harrows are also available.

Considerable weight is put on to the bearings that support the shafts and so it is important that they are regularly greased. Some bearings are of the wood-block type and easily renewed, but on newer models sealed metal bearings are used.

Powered harrow

Powered harrows produce a tilth much more quickly than the conven-

Fig. 12.12   A set of disc harrows in action.
By permission of Kubota Tractors.

tional unpowered harrow. They consist of two or more bars fitted with straight tines. The bars which reciprocate from side to side are driven by the tractor power take-off through a gearbox. The action of the reciprocating tines breaks up the clods of soil, and the fineness of the tilth produced depends on the speed of the tines in relation to the forward speed of the tractor.

Powered harrows are becoming widely used due to the fact that they can 'force' a tilth when it would be unsuitable to use other types of cultivation equipment (Fig. 12.13).

Some powered harrows have a crumbler roller behind them to break up the soil even more and so further reduce the number of operations required to produce a tilth. This has the advantage that it saves time and also helps to conserve moisture by moving the soil the least number of times.

Rollers

Rollers are used to consolidate the surface layers of the soil and also to break up clods. They are also usually used prior to seeding so that a level seed-bed is obtained. Precision seeders require a level seed-bed to work satisfactorily. There are two types of roller:

1. *the flat roller* is used for levelling grassland. It consists of two or more cylindrical cast-iron sections. The longer the roller the more sections are necessary to facilitate turning without pushing up heaps of soil as they slew round;

2. *the Cambridge or ring roller* consists of a series of cast-iron rings mounted on a shaft. The ring roller tends to have a better

Fig. 12.13 A power harrow in action.
By permission of Vicon Ltd.

clod-crushing effect and follows the ground contours better because each ring is separately and loosely mounted.

Rollers are usually sold according to length and weight. Most rollers have some provision for adding weight with either sand or water (ballasting), or in some cases concrete is used for permanent ballast. A larger-diameter roller has a better crushing effect than a smaller roller. The bearings at the end of the shaft require greasing daily during use.

Row-crop cultivators

Many horticultural crops are grown in rows so inter-row cultivations are necessary throughout the growing season. The use of row-crop cultivators requires accurate driving by the operator. The cultivator consists of a frame connected to the three-point linkage of the tractor. Some cultivators are mid-mounted to enable the driver to see more clearly what he is doing. Hoe blades are fastened to the frame and can be positioned so that they cultivate the soil between the rows and work as close to the plants as possible without causing damage.

Two common types of hoe blades are used, A hoes and L hoes. The A hoes are positioned to work down the centre of the inter-row spaces cultivating the soil and cutting through any deep-rooted weeds. The L hoes follow behind the A hoes 15 mm shallower and move the soil between the rows of plants. This disturbs any small weeds which may be growing. On the first hoeing discs should be used in front of the L hoes so that the plants are not disturbed when the very fine seed-bed, which has probably capped, is cut close to the plants. The discs are

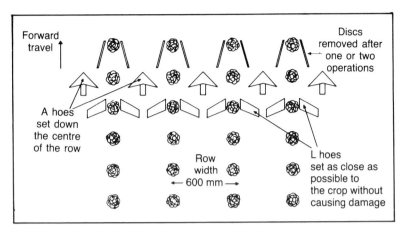

Fig. 12.14   A typical setting of a row-crop cultivator.

usually angled so that the plants are left on a small ridge which makes hand singling easier. After one or two operations these discs should be removed or else damage to the plants will result. The setting up of a tool bar and hoes must be done with care. A typical setting is illustrated in Fig. 12.14.

It is important that the number of rows hoed at one time is the same as the number drilled at one pass so that the tractor wheels can run in the same marks and prevent undue consolidation. The frame must be levelled in all directions by using the right-hand-side lift rod and top link. An easily adjusted tool bar is an advantage as final positioning of the hoes is really only possible in the field. Some machines have spring-loaded hoes which enable them to follow the land contours. Depth of cultivation is 50 mm and this is controlled by two land wheels.

Steerage hoes enable much closer hoeing of the plants to be obtained. Rear-mounted steerage hoes require an operator on the hoe, but mid- or front-mounted hoes can be steered by the tractor driver. Accurate steering is very important and forward speed should be suited to the crop. If the hoe has been set correctly, only one row need be watched, but an occasional glance at the other hoes will spot blockage due to weeds and stones.

Maintenance of row-crop equipment is confined to daily greasing of all moving parts, i.e. land wheels and steering mechanism, and tightening of hoe blades.

Field use of cultivating equipment

The correct use of cultivation equipment when preparing a seed-bed is very important. The sequence in which the implements are used is determined by soil type, time of year, and type of seed-bed required. Ploughing is usually carried out in the autumn so that the furrows can be left to be broken down by frost action during the winter. On some light soils spring ploughing can be satisfactorily carried out.

Cultivation of the furrow slice commences in the spring as soon as the soil has dried out to allow satisfactory breakdown of the clods. When preparing a seed-bed it is usual to start shallow and work deeper so that unweathered soil is not brought to the surface. After ploughing, either the rigid-tine cultivator or heavy harrow can be used to break down the soil. In some cases disc harrows break down the soil while a series of rotovations at different speeds can produce a reasonable tilth. Following the rigid-tine cultivator, either medium harrows or spring-tine harrows break down the soil even further. Alternatively, the use of a powered harrow can reduce the number of operations required. Finally the roller is used to obtained a level seed-bed so that satisfactory seeding can take place.

To illustrate this sequence a typical situation is considered where a crop of cabbages is to be grown on heavy soil.

1. Ploughing 200 to 250 mm deep in the autumn.
2. The soil is then left for frost action to take place.
3. In early spring the rigid-tine cultivator is taken across the furrows.
4. This is followed by two harrowings in each direction.
5. The spring-tine harrow could then be used if a fine tilth has not been obtained.
6. Finally, one or possibly two rollings followed by transplanting.

Experience plays a large part in the correct use of cultivating equipment as there are so many factors involved which can vary from year to year.

Questions

1. What is the function of a plough?
2. Make a drawing to show the soil-working parts of a plough.
3. Name the three types of plough body in common use.
4. When should each type of body be used?
5. What effect does each type of body have on the furrow slice?
6. List the methods of adjusting the tilth produced by a rotary cultivator.
7. List the uses of a rotary cultivator.
8. Draw the soil-working parts of the three common cultivators.
9. Why are harrows used?
10. What advantages does the powered harrow have over conventional cultivation equipment?
11. Make a drawing to show how the hoes on an inter-row cultivator must be positioned to work in a crop with row spacings of 450 mm.
12. Place the following implements in sequence of use when preparing a seed-bed; harrows, cultivator, plough, roller, spring-tine cultivator.
13. Describe the sequence of cultivation operations for a crop of carrots on a medium loam.

13 Fertilizer distributors and seed-drills

Fertilizer distributors

Fertilizers are widely used in horticulture to provide plants with the foods they require for economic growth. Distributors are used in order that fertilizers can be spread evenly over large areas so that each plant gets an equal share.

Good fertilizer distributors should meet the following requirements and these should be considered when selecting a new machine:

1. they should be able to spread both granular and powdered fertilizers at a wide range of application rates, 17 to 335 g/m^2 (grams per square metre);
2. the application rate should be easily adjustable;
3. they should be capable of applying top-dressings to lawns and sowing grass seed;
4. they should be easily dismantled for cleaning and designed to reduce corrosion by the use of plastics and other non-corroding materials.

There are five common types of distributor used in horticulture:

1. the conveyor and brush;
2. the roller feed;
3. the spinning disc;
4. the oscillating spout;
5. the notched roller.

Conveyor-and-brush distributors

This type of distributor can be either hand-pushed for use on lawns or tractor mounted for use on larger areas (Fig. 13.1). It consists of a hopper whose bottom takes the form of a rubber conveyor belt. The conveyor belt is driven by land wheels and as it rotates it carries the fertilizer towards the front or rear of the hopper. The fertilizer passes out through a gap between the bottom of the hopper and the conveyor belt and is spread by a rotating brush to ensure even application. Some machines do not have a brush and the fertilizer falls off the conveyor on to the ground. Drive to the conveyor is usually by gears from the land wheels, and the conveyor speed can be altered by changing the gear ratios. Other makes use the wheel axle as the drive roller for the belt.

Application rates on these machines can be altered by the following methods:

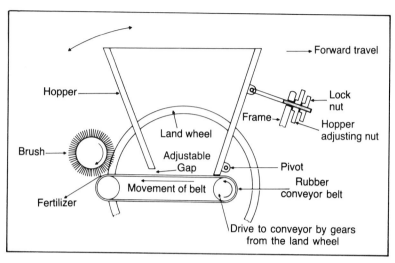

Fig. 13.1   The conveyor-and-brush distributor.

1. adjusting the distance between the bottom of the hopper and the conveyor either by raising the whole hopper or by raising a slide which alters this gap. If widened more fertilizer can pass beneath it and the application rate is increased;
2. where a gear drive is provided, changing the gear ratios will increase or decrease the speed of the conveyor and thus the application rate.

However, it should be noted that the same setting of the hopper will give different application rates for granular and powdered fertilizers. Although the instruction book for the machine may provide a guide to the application rate, the exact setting can only be determined by a calibration.

Calibration procedure

To calibrate a small pedestrian-operated machine 900 to 1 000 mm wide, a square with sides 1 m long should be marked on a clean floor. Next the distributor is filled with fertilizer and is pushed over the square starting 1 m from it so that the machine is working properly by the time it has reached the square. The fertilizer in the square is swept up and weighed. The weight collected will be the application rate in grams per square metre that the distributor is applying. If it is not the rate required then the distributor must be adjusted and recalibrated. This must be repeated until the required application rate is obtained.

Roller-feed distributors

This type of distributor is either tractor mounted or trailed. It consists of a hopper, the bottom of which is formed by two long neoprene rollers (Fig. 13.2). The two rollers are driven independently and rotate in opposite directions. The fertilizer is carried between the two rollers and

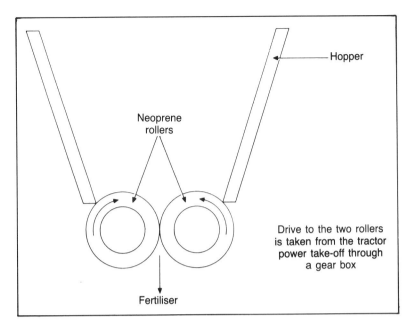

Fig. 13.2   The roller-feed distributor.

because of the spongy nature of the neoprene is not damaged. The speed of rotation of the rollers controls the flow of the fertilizer and hence the application rate. The drive to the roller is usually from the tractor power take-off and the speed is reduced by means of a gearbox which allows different ratios to be selected to give different application rates. The action of this type of distributor is more positive than the earlier types and enables accurate application rates to be obtained. It is also suitable for sowing seeds and the cushioning effect of the rollers prevent damage to the seeds.

For calibration the makers supply special trays with the machine which can be fastened under the rollers. The distributor mechanism is operated for a distance of 20 m using the tractor gear which will be used for spreading. The fertilizer collected in the trays is tipped into a special calibrated container and the rate of application is obtained by comparing the level of fertilizer with a scale on the container.

Because of the simplicity of the distributor and the use of neoprene rollers, corrosion is not a problem.

Spinning-disc distributors
This type of distributor is either tractor mounted or trailed and is more suitable for larger areas (Fig. 13.3). For smaller areas a pedestrian-operated distributor is available. The fertilizer is contained in either an inverted conical or larger trough-shaped hopper. It is spread in a wide arc by a horizontal rotating disc or discs driven at high speed from the tractor power take-off or from the land wheels of trailer or pedestrian-

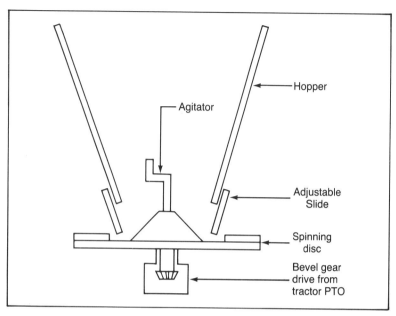

Fig. 13.3   A spinning-disc distributor.

operated models. The fertilizer is fed on to the centre of the disc by gravity or by a conveyor chain from the hopper above the disc. This type of distributor needs an experienced operator if even application rates are to be achieved as less fertilizer is applied towards the sides of the area spread and so overlapping is necessary. Too big an overlap will cause excessive application while insufficient overlap will give less than the required rate. The manufacturers provide details in the handbook about what is called the 'effective spreading width' and the operator can easily determine how far apart adjacent runs should be. Variations in application rates are also experienced when using granular or powdered fertilizers. Care must be taken when pushing the pedestrian-operated model as too much variation in forward speed will affect the spreading width.

Application rates on these machines can be altered by:

1. adjusting the size of the opening in the hopper through which the fertilizer falls on to the plate. Increasing the size of the opening increases the application rate;
2. adjusting the forward speed of the tractor. A slower speed will give an increased application rate. The tractor engine should be set to give the standard power-take-off speed.

Calibration procedure

Calibration of this type of distributor is not as easy as for the other types. One method is as follows. A number of polythene sheets 1 m square are laid on the ground side by side – opened fertilizer bags are suitable for this – and a trial run is made over them. The fertilizer on the squares is collected and weighed and the average weight on each square is the approximate application rate per square metre. If it is not the rate required the distributor can be adjusted and recalibrated. Egg trays can provide a useful alternative to polythene sheets for calibration. If a greater amount of fertilizer is found on one sheet than on the others it can indicate uneven spreading from left to right.

Oscillating-spout distributors

The performance of this machine is similar to the spinning-disc distributor in that it spreads an area much wider than the machine itself, but instead of the fertilizer being thrown from a high-speed rotating disc, it is thrown from a spout oscillating in a horizontal plane. As the spout moves backwards and forwards it throws the fertilizer out at a high speed and gives a rectangular spreading pattern. The distributor has a conical hopper with holes in the bottom which directs the fertilizer on to a regulating disc. An agitator is provided to prevent the fertilizer bridging. In the regulating disc are a series of arrow-shaped holes whose size can be varied by means of slides. The fertilizer passes through these holes into the oscillating spout. The wider open the holes, the more fertilizer can pass through them and so the application rate is increased. The size of the holes is controlled by a lever on a quadrant which can be reached from the tractor seat. This has numerous positions ranging from wide open to fully closed, and is used to cut off the fertilizer when turning at the headlands (Figs. 13.4 and 13.5).

The range of application rates of fertilizer is from 22 to 2 700 kg/ha (kilograms per hectare).

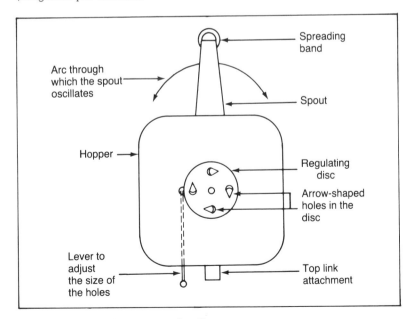

Fig. 13.4   An oscillating-spout distributor.

Fig. 13.5   An oscillating-spout distributor in action.
By permission of Vicon Ltd.

The use of plastic and stainless steel on this distributor help to prevent corrosion.

As with the spinning-disc distributor, care is needed when using this type of machine to ensure that the correct overlap is given so that even application is obtained. The effective spreading width can be obtained from the manufacturer's instruction book as it differs depending on the type and rate of fertilizer used. The procedure for calibrating this machine is similar to that used for the spinning-disc distributor. Another method is to remove the spout, run the machine with the tractor stationary, and collect the fertilizer in a bucket. If this is done over a period of time and the fertilizer weighed, the application rate can be calculated by referring to the manufacturer's instruction book.

Notched-roller distributors
This type of distributor is usually available as a small machine for the domestic market. It consists of a hopper whose bottom is formed by a wooden or plastic roller. In the roller there are a series of notches and the fertilizer is carried round in these notches between the bottom of the side of the hopper and the roller. The roller is driven by the land wheel at each end. The application rate can be varied by either replacing the roller with one having either more or less notches in it or by using a variable-sized outlet slot. This type of distributor is calibrated in the same way as the conveyor and brush type (Fig. 13.6).

Corrosion control
Corrosion of distributors can occur very easily due to the chemical action of the fertilizers when they become damp. This can result in the

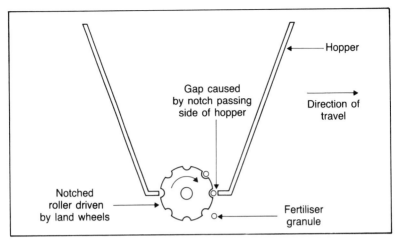

Fig. 13.6   A notched-roller distributor.

annual replacement of parts which is an unnecessary cost. A wide range of distributors today use plastics, rubber, and fibreglass to try to reduce corrosion. It can be kept to a minimum if the following precautions are observed:

1. all traces of fertilizer should be cleaned from the machine after use;
2. fertilizer should not be left in the machine overnight;
3. the distributor should not be left outside overnight or when not in use;
4. at the end of the season the working parts should be stripped down and cleaned thoroughly, removing all traces of fertilizer;
5. after cleaning all the parts should be painted with waste oil and any parts where the fertilizer has worn away the paint should be repainted;
6. finally the distributor should be covered with sacks to prevent too much dust sticking to the waste oil when in store;
7. if any parts are worn they should be ordered and replaced so that the distributor is ready when it is next needed.

Seed-drills

Where a large area of land is to be sown it is more easily and evenly done with a mechanical seed-drill. Seed can be sown in two ways:

1. By broadcasting, where the seed is sown evenly over the whole area.
2. In rows where the seed is sown in evenly spaced rows over the area.

Broadcasting can be done either by hand, with a fertilizer distributor or with a broadcasting drill. This is similar to the random seeder described below, but without coulters so that the seed falls indiscriminately on the surface of the ground.

Most horticultural crops are grown in rows so that operations on the growing crop can be carried out more easily. A simple seed-drill has a coulter which opens a slot in the soil; a metering device which meters seed at the correct rate into the slot; a means of covering the seed; and a means of pressing the soil round it to provide good conditions for germination and growth.

A simple pedestrian-operated drill may consist of one such unit pushed along by hand, but several units can be fitted to a tool bar which can be mounted on the hydraulic-lift system of a tractor. Each unit is independently mounted so that it follows ground surface variations.

Chief parts

The coulter is boat-shaped and is forced through the soil to form the slot into which the seeds are delivered. Covering is done by a chain dragged over the slot or by a pair of blades set at an angle to pull the soil into the slot. Two land wheels are fitted to each unit, one at the back and one at the front, the rear one pressing the soil into close contact with the seed.

Metering device

The most common type of seed-metering device is known as the random seeder. It will sow different sizes of seed in rows at varying depths according to the type of seed (Fig. 13.7).

The seed is placed in a steel hopper at the bottom of which there is an opening. Beneath the opening there is a disc in which there are a series of holes of varying diameters. The holes are designed so that one size can be selected to suit the type of seed to be sown and this is positioned under the opening in the hopper. The holes have either numbers or letters to identify their size and must enable one or more seeds to pass through them easily. A constant supply of seed to the opening is provided by either an agitator or a rotating brush driven by

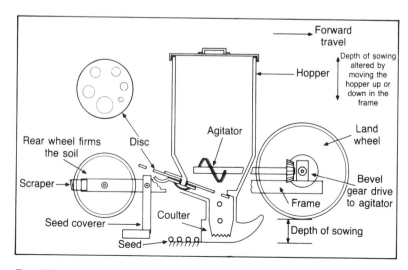

Fig. 13.7 A random seeder.

one of the land wheels. A shutter operated by a trip mechanism shuts off the flow of seeds through the hole when the unit is lifted from the ground. This type of seeder sows a continuous line of seeds which may require thinning or singling after germination.

Adjustments

1. Depth of sowing is varied by moving the coulter up or down in relation to the land wheels. The land wheels are kept firmly on the ground by an adjustable spring-tensioning device between the unit and the frame of the tool bar.
2. Changing the size of the hole in the disc enables different types of seed to be sown.
3. The rate of sowing is controlled by the size of hole in the disc. A larger hole enables more seed to leave the hopper so the sowing rate is increased. The manufacturer provides tables so that the correct hole can be selected for the required sowing rate.
4. The speed of operation of this type of seeder should not exceed 5 km/h (kilometres per hour) (3 mph), as higher speeds will cause erratic seeding.

Precision seeders (spacing drills)

Over recent years there has been a rise in the cost of labour and seed. Some attempt to reduce the cost of unwanted plants has resulted in the development of the precision seeder. A precision seeder sows single seeds at predetermined spacings so reducing the quantity of seed required, although as the germination of every seed cannot be guaranteed seeds must be sown more frequently than finally required to allow for losses.

Apart from the metering mechanism the unit is the same as that described earlier. The metering mechanism is usually one of two types:

1. a rubber belt with a series of holes along its length (Fig. 13.8);
2. an aluminium cell wheel with a series of holes around its circumference (Fig. 13.9).

As the belt or cell wheel rotates underneath the hopper, one seed drops into each hole. To prevent other seeds not completely in the hole from leaving the chamber, a repeller wheel rotates against the belt or cell wheel. The seed is carried by the belt or wheel to a point where it can drop out into the groove in the soil made by the coulter.

Correctly graded seed must be used in a precision seeder so that there is room for only one seed in each hole and seeds do not project above the hole resulting in damage.

Drive to the mechanism is either by vee belt from the land wheel of each unit or by gears and shafts from a master land wheel. Like the random seeder, high speed of operation can cause seeds to be missed and the recommended speed is 3 km/h (2 mph).

Seed spacing and the type of seed sown is altered by changing the seed belt or cell wheel for one that has differently spaced holes or different-sized holes for other seeds.

Operation

Seeder units are positioned along the tool bar according to the row

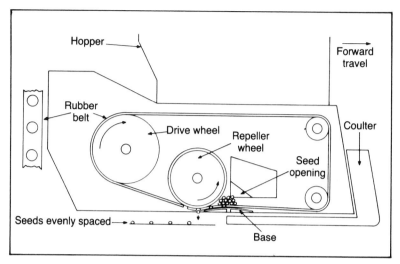

Fig. 13.8   A rubber-belt precision seeder.

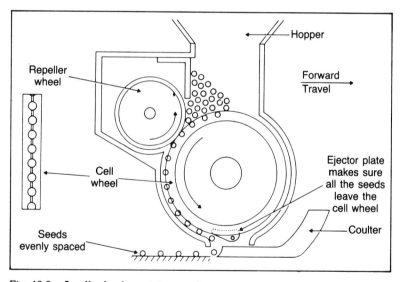

Fig. 13.9   A cell-wheel precision seeder.

spacing required. Also, the wheel settings of the tractor must suit the row spacing as the wheel must run between the rows or otherwise the drill units will run in the wheel markings. Both front and rear wheels must be set at the same width. Most tool bars are provided with markers so that accurate positioning of the units can be made on the return run.

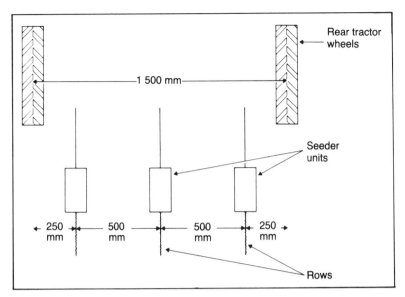

Fig. 13.10   The setting of a 3 unit drill sowing in 500 mm rows.

Figure 13.10 illustrates the correct setting for a 3 unit drill sowing in 500 mm rows.

Most units are mounted on the tractor three-point linkage. The top link and the right-hand side levelling box should be used to level the tool bar. Stabilizers should be fitted to prevent excessive side swing.

Summary

For successful field operation the following points should be observed.

1. the units and the tractor wheels should be set for the desired row width;
2. all moving parts should be lubricated daily according to the manufacturer's instructions;
3. a check should be made to see that the correct belt or cell wheel is being used;
4. each seeder unit should be turned over by hand before lowering to check that it is sowing;
5. the forward speed must not exceed 3 to 5 km/h (2 to 3 mph);
6. the units should be kept dry and a check made to see that no seeds become lodged in the drive mechanisms;
7. when sowing is finished all traces of seed must be removed from the hopper and seed-chamber;
8. all the seed-belts or cell wheels should be removed from the units and stored separately.

In order to reduce the germination of weeds which may compete with the young root seedlings band spraying is often practised. A sprayer is mounted on top of the tool bar with a special nozzle fastened behind each of the seeder units. After the seed has been sown the soil is sprayed with a pre-emergent weedkiller which kills the weed seedlings as they germinate, but whose toxic effect has gone by the time the root seedlings germinate.

Questions

1. Draw the fertilizer metering mechanism for three types of distributor.
2. How would you calibrate a small pedestrian-operated fertilizer distributor 900 mm wide?
3. Why is thorough cleaning of fertilizer distributors essential?
4. Outline a typical cleaning procedure.
5. List the main components of a seed-drill.
6. Draw the metering mechanism for one type of precision seeder.
7. Why must graded seed be used in precision seeders?
8. Draw the layout of five drill units on a tool bar so that they can be used to sow a crop in 600 mm rows.

14 Nursery machinery

Planters

Although many horticultural seeds can be sown directly into the ground where the plants will develop and mature, others are sown in boxes, nursery beds, or in soil blocks with the intention of planting-on the young seedlings into their permanent positions when they are strong enough to endure the move. Planters which are designed to do this and can set out plants, soil blocks, or pot balls do so only semi-automatically; although the plants are planted mechanically at the required spacings they have to be fed into the machine individually in the correct position by hand.

Planters consist of up to seven individual planter units mounted on a frame which can be carried by the hydraulic-lift system of a tractor. Each unit consists of a coulter which forms a slot in the soil; a planting mechanism which places plants correctly into this slot at the required spacing; a pair of press wheels set at an angle to each other to consolidate the soil around the roots; and a seat to carry the operator feeding the plants into the planting mechanism (Figs. 14.1 and 14.2).

Construction

The coulter is not unlike a larger version of the coulter on a seeder but has two rearward-projecting side plates to hold open the slot it forms and to prevent the soil flowing back into it until the plant has been placed in position.

One important type of planting mechanism consists of two flexible steel discs each mounted at the end of a short shaft and at an angle to each other so that the discs are pressed lightly together over almost half their circumference. The discs are positioned vertically and are gear driven from the press wheels. The operator can insert a plant in the gap between the discs at the top of their revolution with its roots protruding upwards. As the discs turn they come together, lightly gripping the plant, its leaves lying between the discs where they are protected, and carry it around until it is held with its roots in the slot in the soil formed by the coulter. As the machine moves forward the soil flows around the roots, and the discs are parted as they continue their revolution, releasing the plant (Fig. 14.1).

Plastic markers can be bolted to one of the discs to indicate to the operator where each plant should be inserted. Soil blocks may be set by replacing the markers with grippers fitted with claw-like forks which hold the blocks as the discs rotate.

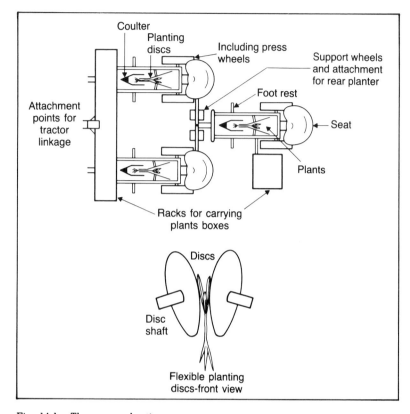

Fig. 14.1   Three-row planting – narrow rows.

Instead of steel discs, rubber ones held together for part of their revolution by metal rollers can be used. Notches moulded around the edges of the discs help the operator to space the plants correctly, and ridges moulded on the insides of the discs prevent the plants from slipping.

More precise plant spacing may be possible with a planting mechanism in which the operator places the plants individually in grooved grippers attached at equal intervals. around the circumference of a vertical disc. Each plant is held in its gripper by a cushion pad which closes on the stem and is lightly held against it by a spring. When the disc has rotated, and the roots of the plant are in the groove in the soil made by the coulter, the cushion pad is mechanically pulled away, releasing the stem.

Plant spacing can be altered by changing the number of grippers around the disc or by altering the disc speed; weaker springs can be fitted for soft-stemmed plants.

When plants such as tree seedlings which are too long to fit into the mechanism have to be planted the mechanism can be dispensed with

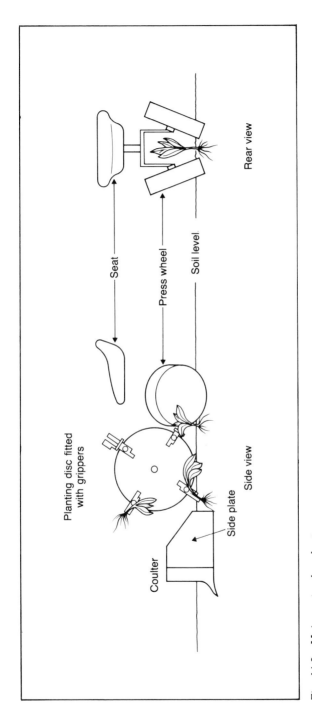

Planting disc fitted with grippers

Seat

Press wheel

Soil level

Rear view

Coulter

Side plate

Side view

Fig. 14.2 Main parts of a planter.

and the plants inserted by hand at the right spacing into the slot formed by the coulter and held there until the soil flows back around them.

Two wide-rimmed press wheels, also at an angle to each other, compress the soil around the roots and carry much of the weight of the unit and its operator, although additional wheels may be fitted. A gap of 15 to 40 mm is left between the wheels so that they can pass on either side of the plant without damaging it. When soil blocks are being set, the wheels need to be further apart. The seat and foot rests are fitted as low as possible so that the operator does not have to bend far when feeding the planting mechanism. Racks are provided on the main frame of the machine to carry boxes of plants.

There are also provisions for fertilizing and watering attachments. Both the fertilizer and the water can be placed in the slot so that they are close to the plant roots, and the soil infilling the slot prevents rapid evaporation of the water. Instead of applying water continuously, a mechanism can be fitted to apply a quantity of water to each individual plant. This reduces the weight of water that needs to be carried.

Operation

1. Planter units are positioned along the frame according to the row spacing required. The units are too wide to position side by side when narrow row spacings are required, but row widths down to 300 mm can be obtained by fitting the units in two rows staggered one behind the other.
2. The description of markers, stabilizers, general settings and the setting of the wheels of the tractor given in Chapter 13 (p. 105) also apply to planters.
3. Spacing within the rows can be altered by varying the number of plastic markers or grippers fitted to the disc, or by changing or transposing gears in the gear drive to the planting mechanism.
4. Coulter depth is adjustable, and it is set so that it is just deeper than the longest roots of the plants to be set.
5. Speed of work is limited by the rate at which the operators can feed the planting mechanisms. When very close planting is required very low forward speeds are necessary. These may only be obtainable by modifying the transmission of the tractor to include an extension giving an additional ratio which enables extra low forward speeds at near full power to be obtained.

Maintenance

Regular lubrication and other simple checks are required in accordance with the manufacturer's recommendations.

Undercutters

Undercutters are used when lifting horticultural subjects ranging from shrubs and trees to root vegetables.

In its simplest form it consists of a straight or slightly vee-shaped blade which is pulled horizontally through the soil at a predetermined depth under the subjects to be lifted, cutting any deep taproots that they may have. Backward-projecting fingers attached to the cutter may help to

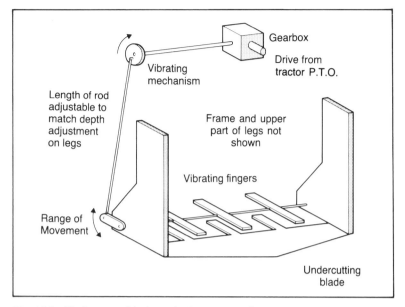

Fig. 14.3   Undercutter blade and vibrating mechanism.

separate the plant from the soil. On more elaborate lifters this is done by vibrating the share and attached fingers, or by mounting the fingers separately from the blade on their own shaft so that they can be vibrated independently of the blade which is fixed. Normally, the two ends of the blade are supported by legs attached to a frame mounted on the hydraulic-lifting system of a tractor, and the length of these legs can be adjusted to undercut at the depth required. The fingers are vibrated by a crank mechanism driven from the power take-off of the tractor, and the severity of vibration can be varied (Fig. 14.3).

Usually the lifter is fitted directly behind the tractor so that with its wheels extended it can straddle a nursery bed of up to about 2 m wide. However, subjects too tall to pass under a tractor can be lifted by single-leg machines offset to one side of the tractor, and these can lift semi-mature trees of up to about 5 m in height. The forces exerted on the tractor by such a lifter tend to slew it around, and make it difficult to steer down the row. This is overcome by balancing the one-sided forces exerted with a single cultivating tine mounted on the frame on the other side of the tractor from the lifter. Such machines may require a tractor developing 50 kW or more to operate them.

Compost and soil shredders

There are several types of compost and soil shredder, but most of them operate by disintegrating the material with fast-moving blades until it is sufficiently fine to pass out through some form of sizing screen.

The material to be shredded is shovelled into a hopper at the top of

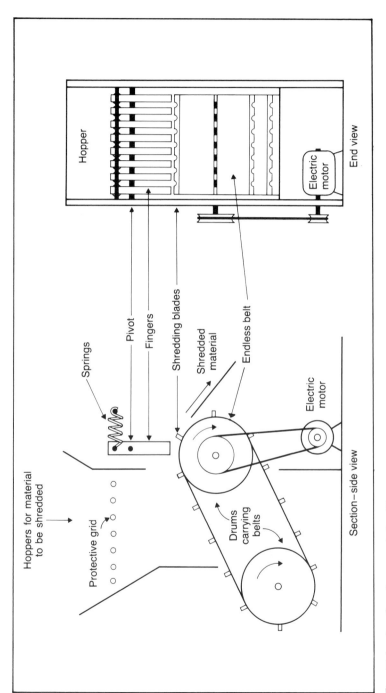

Fig. 14.4   A soil and compost shredder.

End view

Hopper

Electric motor

Section–side view

Springs

Pivot

Fingers

Shredding blades

Shredded material

Endless belt

Hoppers for material to be shredded

Protective grid

Drums carrying belts

Electric motor

the machine, although higher-capacity machines may be loaded by a conveyor or by a tractor and fore-end loader fitted with a bucket. From the hopper the material falls by gravity through a coarse safety mesh into the disintegrating mechanism (Fig. 14.4).

Shredding mechanism

One type of shredder uses a rapidly rotating cylinder, 225 mm in diameter, fitted with ten serrated blades which shred the material until it is fine enough to pass through a grid of spring-loaded tines, 12 mm apart. Stones too large to pass between the spring-loaded fingers are flung back by the tines and out of the other side of the machine.

An alternative type of machine has the serrated blades mounted on an endless belt running on two pulleys instead of a drum. In yet another, the shredded material has to pass out through a fixed screen instead of between fingers.

In some machines large lumps of turf or soil are broken down prior to shredding by rotating hammers mounted above the shredding blades.

Shredders can be driven by electric motors, small engines, or by the power take-off of a tractor. Portable machines driven by electric motors can only be used close to a suitable power point, whereas machines driven by engine or power take-off can be operated anywhere.

Fig. 14.5   A rotary sieve.

Rotary sieves

Rotary sieves are used for separating large quantities of fine material quickly for incorporation in compost mixtures and for top-dressing.

One type consists of a sloping cylinder, open at the ends, made from mesh. The cylinder is rotated by an engine or electric motor to sieve the soil which is fed in at the high end. Fine material passes through the mesh, whereas clods and stones are carried through and pass out of the other end of the drum.

The gearbox and vee-belt speed-reduction drive incorporate an overrun clutch which allows the heavy cylinder to continue rotating on its own and to slow down gradually after the engine or motor has been switched off. This avoids damage to the driving system which a sudden stop might cause.

A simpler type consists of a short, slightly conical, wire-mesh drum open at one end and set at a slight angle, which is driven by an electric motor or small engine. Common mesh sizes are 3, 10, and 14 mm (Fig. 14.5). Shredded soil is fed into the open end of the drum, fine soil passes out through the mesh, and stones and clods are left in the drum.

Wet soil does not sieve well and tends to block the holes in the sieves.

Questions

1. How does the coulter of a planter differ from that of a seed-drill?
2. What adjustments must be made to the press wheels when soil blocks are being set?
3. What limits the rate of work of a planter?
4. How can the plant spacing within a row be varied on a planter?
5. How can the depth of work of an undercutter be changed?
6. How is the amount of vibration varied on an undercutter with vibrating fingers?
7. What is the principle on which compost and soil shredders work?
8. What is the purpose of the spring-loaded fingers on a shredder?

15 Mowing machines

The areas of grass mown by machine can vary from a fine cricket wicket or bowling-green to playing-fields and motorway embankments. The wide selection of machines manufactured to deal with these different mowing requirements can be grouped under five headings:

1. cylinder mowers – including lawn-mowers and gang mowers;
2. rotating-disc mowers;
3. reciprocating-knife mowers;
4. flail mowers;
5. nylon-cord mowers.

Cylinder mowers (Fig. 15.2)

This group includes hand-propelled and self-propelled lawn mowers, and tractor-drawn and tractor-mounted gang mowers.

The cutting principle of the cylinder mower closely resembles that of a pair of scissors. The grass is trapped between two blades, one of which is stationary. The other rotates and is drawn across the stationary blade causing a cutting action.

The two main components are the cylinder and the bottom blade (Fig. 15.1). The cylinder consists of a number of spiral blades fastened to flanges mounted on a central shaft. The blades rotate, and because they are mounted spirally cause a progressive cutting action across the bottom blade.

The number of blades on the cylinder can vary from three to twelve; the more blades there are the more cuts they make per metre and the finer is the cut. The gear ratio between the cylinder and the drive roller or wheel can also affect the number of cuts per metre. Typically, a three- or four-bladed rough-cut mower gives 33 to 40 cuts per metre, a six-bladed lawn mower gives 72 to 82 cuts per metre, a twelve-bladed lawn mower gives 140 to 160 cuts per metre, and a five-bladed gang mower gives 23 to 33 cuts per metre.

Where a very fine cut is required, as on a bowling-green or cricket wicket, then a ten- or twelve-bladed cylinder will give the best result. On the other hand where a coarse cut is acceptable on a motorway embankment, a cylinder with three to six blades will be satisfactory. There are some machines now available in which the speed of the cylinder in relation to forward speed can be adjusted, faster speed giving more cuts per metre and vice versa.

The bottom blade is a flat piece of steel varying in thickness from 3 to 6 mm. The thickness depends on the closeness of cut the mower is

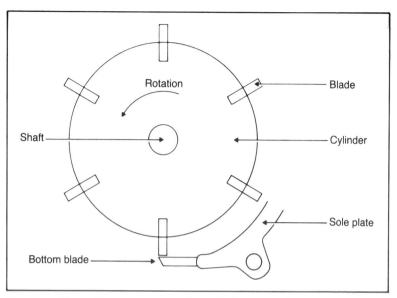

Fig. 15.1   The cutting components of a cylinder mower (side view).

Fig. 15.2   A pedestrian-operated cylinder mower.

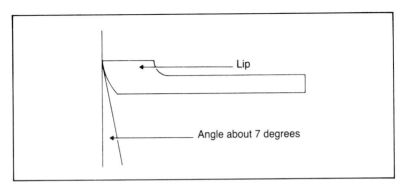

Fig. 15.3 The bottom blade.

designed to give. The leading edge is angled to give a 'keen' edge to cut against. The angle is less than 90 degrees, as shown in Fig. 15.3. With the exception of those on fine-turf mowers, heavy-duty bottom blades are 'lipped' to give them extra strength. Thin blades are used on mowers designed to give close cuts because it allows the cutting edge to get closer to the ground.

The rear roller provides the drive to the machine. The cylinder of a hand mower is chain or gear driven from the rear roller. The drive from the engine of a motor-driven mower is taken to the rear roller and also to the cylinder by chain or belt. The rear roller should be made up of at least two sections, which make it easier to turn the machine without damaging the grass. Some rear rollers are 'ribbed' to improve their driving grip in wet conditions.

The front roller is used to support the front of the mower and to allow height-of-cut adjustment. Good mowers should have removable sections in the front roller to facilitate the cutting of long grass if it has been allowed to get out of hand.

There are three common ways by which cylinder mowers can be driven:

1. *by hand*, although these are restricted to between 250 and 450 mm working widths;
2. *by petrol or diesel engine working on either two- or four-stroke cycles*. Generally the latter proves to be the less troublesome. Width of cut ranges from 300 to 900 mm;
3. *by mains electric*. This machine incorporates a 200 to 250 W heavy-duty electric motor. The disadvantages of this form of power are that there must be a suitable three-pin power socket near at hand and that the distance that it can be operated away from the socket depends on the length of cable. The operator has to follow a planned cutting pattern to avoid the cable which tends to get in the way when cutting. Advantages of this form of power are the cheaper purchase price of the machine, coupled with the low running cost (less than 2p per hour). They are quieter and suitable for use in hospital grounds and schools.

Adjustments to cylinder mowers during the cutting season

1. *Height of cut.* This is done in most cases by adjusting the vertical position of the front roller. Raising the roller reduces the height of cut, lowering it increases the height of cut. It is tested by placing a straight-edge against the underside of the front and rear rollers, so that the distance between the bottom blade and the straight-edge can be measured (Fig. 15.4).

2. *Cutting action of the blades.* On all good mowers there is some method of adjusting the closeness of the blades on the cylinder to the bottom blade. This can be done by either moving the whole cylinder closer to the bottom blade by means of adjusters at each end of the cylinder shaft, or by bringing the bottom blade closer to the cylinder. In this case, at each end of the sole plate, there are screw adjusters which bring the bottom blade either nearer or further away depending on which way the screws are turned (Fig. 15.5). The correct adjustment is checked by placing a piece of paper between the cylinder and the bottom blade, then by turning the cylinder by hand,

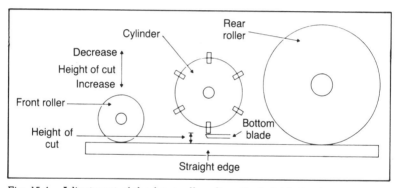

Fig. 15.4   Adjustment of the front roller alters the height of cut.

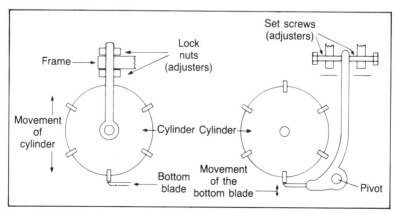

Fig. 15.5   Methods of adjusting the cutting action of the blades.

Fig. 15.6   Checking the correct setting of the blades with a piece of paper.

it should cut the paper cleanly without tearing it (Fig. 15.6). This test must be carried out in at least three places along the bottom blade: in the centre and at both ends, otherwise a cylinder may be correctly set at one end and not at the other.

Maintenance of cylinder mowers

All moving parts should be oiled or greased as recommended by the manufacturer. Too much oil will cause damage to the grass if it drops on to it. A regular check should be made on the machine's cutting performance; if faulty the cylinder clearance should be checked and adjusted as necessary. The action of the cylinder blades rotating against the bottom blade should retain their sharpness, but from time to time the cylinder blades may need regrinding. This is a task which should not be undertaken by the operator unless he has had instruction and uses special grinding equipment. A process known as back-lapping can be used to sharpen cylinders without the need to dismantle the cutting mechanism. A grinding paste can be spread on the cutting edges and the cylinder rotated in reverse against the bottom blade. This will remove small traces of metal and ensure that the surfaces are equally matched. The paste must be washed off with paraffin when the process is complete. All traces of grass must be cleaned from the machine after use by compressed air or a stiff brush. Maintenance of the engine and the drives should be carried out as described in Chapters 5, 6, 7, 8, and 9.

At the end of the season the mower should be thoroughly cleaned, the cutting cylinder slackened off, and the cutting mechanism sprayed with oil. During winter, mowers should be overhauled by a specialist, worn parts replaced, and the cylinder and bottom blade reground if necessary.

Gang mowers

Gang-mowers must also be included with cylinder mowers because their basic components are very similar. The cylinder and the bottom blade are of a much heavier construction than the lawn mower. Gang mowers are either tractor mounted or trailed, but smaller trailed units are now available to use with the large range of small tractors described in Chapter 4. Rear-mounted gangs are usually restricted to 3 units because of their weight which is carried by the hydraulic system of the tractor, but trailed gangs can consist of up to 9 or 11 units. Gang mowing sets based on a tractor with up to 7 mower units mounted at the front, sides, and rear are available (Fig. 15.9). Self-propelled gang mower units are widely used on areas where speed of operation and a high degree of manoeuvrability are required. The fineness of cut depends on the type of cylinder used.

Fig. 15.7   A set of trailed gang mowers being pulled by a small tractor.
By permission of Allen Power Equipment Ltd.

The number of blades on the cylinder governs the type of finish produced and usually varies from four to nine. The cylinder is driven from the land wheels by gears enclosed in a case which acts as an oil-bath. There is a drive arrangement from both wheels, which includes

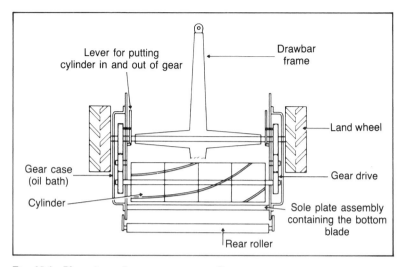

Fig. 15.8 Plan view of a gang mower unit.

an overrun mechanism so that corners can be turned, and a device for disengaging drive so that the cylinder can be put out of gear for towing from place to place. A typical width of a gang mower unit is 750 mm (Fig. 15.8).

The height of cut can be altered by the rear roller when fitted, or by a lever and quadrant on each unit, if there is no rear roller. The rear rollers are usually fitted with some form of damping unit, so that undulations in the ground can be closely followed. The wheels can either be steel with lugs to give good traction, or rubber tyred. The tyres may be either grassland types used where the tread must not show, or lugged where good grip is essential.

Over the last few years there have been changes in the drive systems used on gang mower units. Some models are self-propelled as shown in Figure 15.9 and have a separate belt or shaft drive to each of the units. Each unit is mounted separately so that it can follow undulations in the ground. These self-propelled units are smaller than a tractor-operated set of gang mowers and are very manœvrable.

Larger tractor-operated gangs can be driven hydraulically. (Fig. 15.10). Each unit of the gang has a hydraulic motor mounted at the end of its cylinder and this provides the drive. A reservoir of oil is carried by the tractor and the oil is pressurized by a pump and passed to each individual hydraulic motor through flexible pipes. The oil is returned from each motor to the reservoir by a second pipe.

A further development in gang mower drive is to use the power take-off of the tractor to drive the gang mower units. The units are mounted on a frame which has a pair of wheels to carry it, and each unit is driven by a series of shafts and belts from the power take-off. See Fig. 15.11 for a typical example of a power-take-off-driven set of gang mower units.

Fig. 15.9   A motor triple.
By permission of Ransomes, Sims & Jefferies Ltd.

Fig. 15.10   A set of hydraulically-driven mounted gang mowers.
By permission of Huxleys Grass Machinery.

Fig. 15.11   A set of power-take-off-driven gang mowers.
By permission of Sisis Equipment (Macclesfield) Ltd.

The advantage of these new types of drive arrangements is that they are more positive and will not slip or stop if traction on the land wheels is lost. This enables grass cutting to be carried out under wetter conditions than is possible with land-wheel-driven units. The speed of the cylinder can also be varied in relation to the forward speed of the tractor. A lower forward gear will give more cuts per metre.

Gang arrangements

Figure 15.12 shows some typical ganging arrangements with the areas cut per hour at a speed of 11 km/h (7 mph). Smaller self-propelled units can cover 1 to 1.5 ha/h (hectares per hour).

Maintenance of gang mowers

All grease nipples should be lubricated daily. The oil level in the gear-case should be checked and, if necessary, topped up with the grade of oil recommended by the manufacturer. The oil in the gear-cases should be changed at least once a year. The pneumatic tyres should be inflated to the recommended pressures and both tyre pressures must be the same or an uneven height of cut will result. A check should be made to see that the cylinder clearance is correct and the bottom blade is sharp. All traces of grass must be cleaned from the machine after use, as described earlier (p. 119).

Rotating-disc mowers

The cutting principle of a rotating-disc mower is that a knife-like edge,

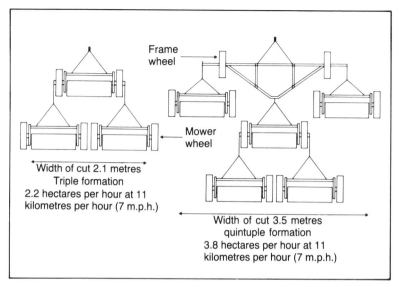

Fig. 15.12 Typical ganging arrangements.

travelling at high speed, cuts through the grass on impact. In practice this is done by having a series of blades, mounted on a disc which rotates at high speed (3 000 rev/min) parallel to the ground (an exception to this is described on p. 127). The number of blades varies from one make of machine to another, but usually there are two, three, or four. The design of the blades can also vary. Some are triangular, others rectangular, but in nearly all cases they can be removed and replaced to expose another cutting edge. The width of cut of these machines is between 350 and 750 mm (Fig. 15.13).

The type of finish produced by the rotating-disc mower is not as good as that produced by a cylinder mower. They are usually used for rough areas such as orchards and verges where the finish is not critical, but some machines available can leave quite an acceptable finish. Most of this type of mower leave the cuttings on the grass, but there are some on the market in which some of the blades are replaced by throwers which throw the cuttings into a grass-box fitted to the machine.

The blades operate at high speed and so must be kept sharp or they will damage the grass by tearing it rather than cutting it. They are made of hardened steel and must be securely fastened on to the disc. On some mowers, the blades have a swinging action and move out of the way if they hit an obstruction. On no account must a mower be used with a blade missing, broken, or clogged with dirt as this will set up excessive vibrations causing damage to the machine. Any operator must, as a matter of course, inspect the blades before cutting and replace or sharpen the blades where necessary. He must also check that swinging blades are free to move.

All rotating-disc mowers have cutting mechanisms that are either engine-powered or driven by an electric motor, but only the larger

Fig. 15.13 A rotating-disc mower with swinging flails.

machines are self-propelled. The engine is usually a small two- or four-stroke petrol engine and on the smaller machines the engine is mounted on its side and the disc is mounted directly on to the crankshaft. In the case of larger machines the drive is often by vee belt from the engine to the disc (Fig. 15.14). Tractor-operated rotating-disc

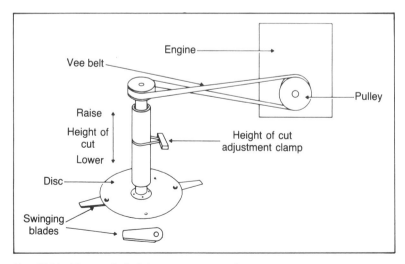

Fig. 15.14 The vee-belt drive on a rotating-disc mower.

mowers are available with one or more rotating discs, giving working widths of up to 6 m.

The height of cut can be adjusted on this type of machine in one of three ways:

1. *by raising or lowering the land wheels.* Raising the wheels lowers the cut and vice versa;
2. *by moving the rotating-disc assembly up and down.* Moving the disc down lowers the cut;
3. *by the use of spacers or washers between the engine and the rotating disc.* Adding washers lowers the cut, removing them raises it.

Hover mower

This type of machine rides on a cushion of air and so does not need any wheels. This makes it very manœuvrable and ideal for bank sides. It is also suitable for cutting wet grass, as there are no wheels to damage

Fig. 15.15   A hover mower.
By permission of Allen Power Equipment Ltd.

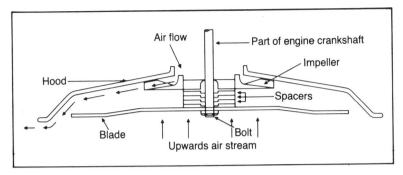

Fig. 15.16 The cutting components of the hover mower.

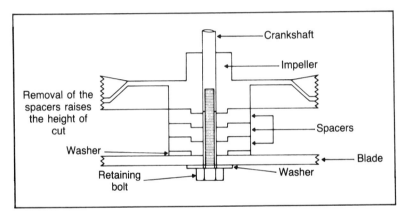

Fig. 15.17 The height of cut adjustment on the hover mower.

the turf. However, it is awkward to move when the engine is stopped (Fig. 15.15).

Mounted on the engine crankshaft is a fan which draws air through holes in the top of a hood. The air is forced down inside the hood and builds up a pressure which lifts the machine off the ground. Also mounted on the crankshaft below the fan is the cutting blade which is a bar with sharpened edges (Fig. 15.16).

To alter the height of cut on this machine 'spacers' are put between the fan and the blade. There are usually three and when all are present the lowest height of cut is obtained (Fig. 15.17).

Maintenance of rotating-disc mowers

All moving parts such as blades and wheels should be oiled daily. All the blades must be sharp and properly secured. Swinging blades must always be free to move and any caked grass must be removed if it prevents their swinging action. Maintenance of engines and air cleaners should be carried out as described in Chapters 6, 7, and 8.

Safety

Rotating-disc mowers can be dangerous if not used properly. The following points, if observed, will help to prevent accidents.

1. Stop the engine and remove the plug lead before making any adjustments to the machine.
2. Never leave the machine with the engine running.
3. Keep hands and feet away from the disc when the engine is running.
4. Always push the mower and never pull it towards you.
5. Remove all visible stones from an area of grass before cutting.
6. Make sure that the rear shield is in place and fitted correctly.
7. Always use the correct type of blade for the machine.
8. Move spectators out of the range of flying stones.

Reciprocating-knife mowers

This type of mower is less popular than the rotating-disc and flail mower. It is suitable for long grass and light scrub in rough areas.

The cutting principle of the reciprocating mower also resembles that of a pair of scissors. The main part of the machine is the cutter bar which can vary in length from 600 to 1 200 mm. This skims the surface of the ground, riding on the back edge most of the time. The cutter bar contains the knife which is made up of a number of triangular knife sections riveted on to a knife back. The knife reciprocates in the cutter bar and is driven at the centre by two lugs on the knife back. The cutter bar also carries the fingers, spaced at 75 mm intervals, and on the inside face of each finger is riveted a hardened piece of steel known as a ledger plate. The sides of the ledger plate are chamfered and it is between this edge and the knife sections that the grass is cut. On some machines the fingers are replaced with a second set of knife sections to reduce the risk of clogging. Cutting is a gradual action due to the fact that the sections are triangular.

The knife sections are held against the ledger plate by knife 'clips' or 'pads' which prevent them from riding up if thick material is being cut. The clips must be set so that they press lightly on the knife sections as they move from side to side. A wearing plate is sometimes provided against which the knife back moves. This must be set so that there is only a slight forward movement of the knife, but is not too tight to resist its side-to-side movement (Figs. 15.18 and 15.19).

Reciprocating-knife mowers are engine driven and are also self-

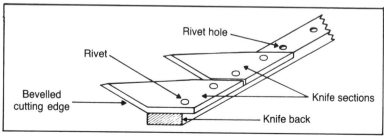

Fig. 15.18   Triangular knife sections riveted to the knife back.

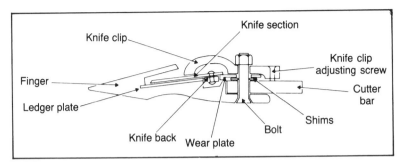

Fig. 15.19   A cross-section of a cutter bar.

propelled. Usually the cutting mechanism operates as soon as the engine is started, but forward movement is controlled by a hand-operated clutch.

Each wheel can incorporate a pawl and ratchet mechanism which acts as a differential for turning corners, but may be locked to prevent the mower from turning when working on steep banks.

Height of cut is controlled by the operator who can press down on the handles to raise the cutter bar.

Maintenance of reciprocating knife mowers

Maintenance of engines should be carried out as described in Chapters 6, 7, and 8. The knife sections must be kept sharp. To sharpen them remove the knife from the machine and file or grind the knife sections, taking care to retain the original cutting angle. Broken or badly worn sections should be replaced. All the finger points should be kept in line. This can be checked by using a piece of string across the points and repositioning any fingers out of line. The edges of the ledger plates should be kept sharp and at the correct angle. The wearing plates must be adjusted to take up any wear on the knife back, and a check made that the knife 'clips' are set correctly. The cutter bar does not need oiling as the sap from the grasses provides lubrication to the moving parts. Oil encourages dirt to stick and increases wear on the cutter bar.

Flail mowers

The flail mower is well established as a horticultural mower (Fig. 15.20). The cutting principle of this machine is similar to that of the rotating-disc mower. A horizontal rotor carrying a number of swinging flails turns at a high speed (3 000 to 4 000 rev/min). The flails have a cutting edge, which is not necessarily sharpened, but cuts the grass on impact. They swing out into their cutting position by centrifugal force (Fig. 15.21).

This type of mower is very suitable for coarse grass or light scrub and is faster than the reciprocating-knife mower and the cutting mechanism has a lower maintenance requirement. Flail mowers will deal with grass up to 1 m high without trouble, leaving the grass as a mulch on the ground.

Machines vary from 750 mm wide, self-propelled pedestrian-operated

Fig. 15.20 A pedestrian-operated flail mower.
By permission of Turner International (Engineering) Ltd.

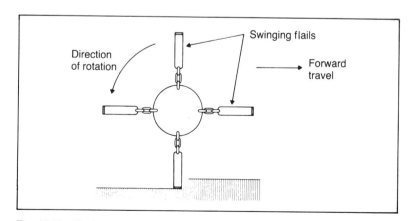

Fig. 15.21 Flail action.

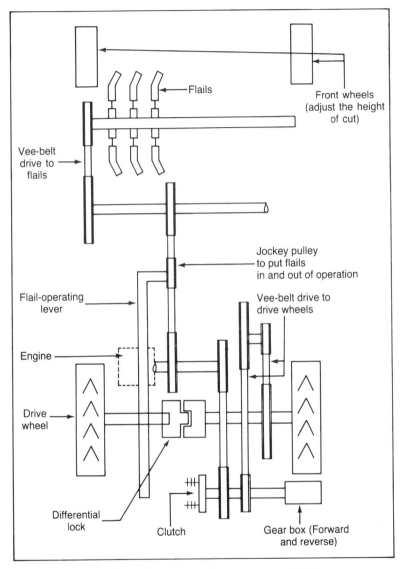

Fig. 15.22 Layout of the drive on a pedestrian-operated flail mower.

mowers to 2.3 m wide tractor-mounted ones.

Pedestrian-operated machines have a 6 to 7.5 kW engine which drives the flails by means of vee belts and the land wheels by chain and sprocket through a gearbox (Fig. 15.22). Most machines have either two or three forward gears and reverse. Hydrostatic transmission can be used on pedestrian-operated machines. This gives infinitely variable

Fig. 15.23 Tractor-operated flail mower.
By permission of Turner International (Engineering) Ltd.

forward and reverse speeds. Where the grass to be cut is on a steep bank a differential lock is an advantage to help prevent the machine sliding down the bank. A two-stroke engine is better than a four-stroke for this work as there is no risk of the engine seizing up due to lack of lubrication. The height of cut is controlled by the front wheels which can be moved up and down, regulating the distance between the tip of the flails and the ground.

Tractor-mounted flail mowers vary from machines mounted at the rear on the tractor hydraulic system and driven off the power take-off, to side-mounted machines on hydraulically positioned arms with a hydraulic motor driving the flails. The latter are very adaptable with reaches of over 8 m either up or down. Double-acting rams are used to position the head and they are ideal for use on motorway embankments (Fig. 15.23).

The flails are not usually damaged when they hit stones because the swinging action allows them to move out of the way on impact. Forward speed can be up to 8 km/h (5 mph). Some makes of flail mower can also be used for cutting hedges. When cutting roadside verges or hedges the operator must place warning signs to inform passing motorists well in advance of the work.

Maintenance of flail mowers

Maintenance of the engine should be carried out as described in Chapters 6, 7, and 8. Before use, a check must be made to see that all the flails are present, none are broken, and if of the swinging type can move freely. Broken or missing flails will cause vibrations which could damage the machine. A check must be made to see that all the drive belts are at the correct tension as specified by the manufacturer.

Fig. 15.24   A nylon-cord mower in action.

Nylon-cord mowers

Pedestrian-operated nylon-cord mowers are now widely used for mowing in tight corners where it is difficult to get other types of mower. They are particularly useful around kerbstones in a cemetery or around the base of trees. A piece of nylon cord is rotated at high speed and this will pulverize grass and weeds on contact (Fig. 15.24).

Power can be supplied to the cord by 12 V battery, mains electricity, or small two-stroke engine. The model chosen will depend on the availability of a convenient power source. Continual use close to brick or concrete will quickly shorten the length of the cord and so frequent lengthening of the cord will be required. This is done by either stopping the machine and drawing out a new length of cord or, on some machines, knocking the reel on the ground will allow more cord to be released.

Safety

1. Always wear safety footwear when using a nylon-cord mower.
2. Clear goggles should be worn to prevent grass and dirt getting into the operator's eye.
3. Always stop the machine by switching off or stopping the engine before lengthening the cord.

Maintenance of nylon-cord mowers

Checks should be made to see that all the guards are in place.
   Grass will need to be cleaned from around the reel.

A new reel of cord will need to be fitted when the cord is used.
All mains electrically operated machines should be correctly earthed unless they are double insulated.

Questions

1. Name four common types of mowing machine.
2. Under what conditions would each be used?
3. What advantages has a twelve-bladed cylinder over a five-bladed cylinder?
4. List the common methods of powering lawn mowers.
5. Describe the method for adjusting the cutting mechanism on a cylinder mower.
6. What area can be cut per hour with five gang mower units travelling at 11 km/h (7 mph)?
7. What methods are used to adjust the height of cut of a rotating-disc mower?
8. List the main safety points to consider when using a rotating-disc mower.
9. Describe the cutting principle of a flail mower.
10. Why should the flails be checked each time the flail mower is used?
11. What maintenance is necessary for the cutting mechanism on (a) cylinder mowers, (b) rotating-disc mowers'.
12. What is a nylon-cord mower most suitable for?

16 Sprayers

Spray chemicals are widely used in horticulture for the control of weeds, pests, and diseases. Accurate and even distribution of the chemical is important because an excessive application may harm the plants, whereas an inadequate application may be ineffective and waste the often very expensive chemical. The range of droplet sizes produced by the sprayer is also important, as are other factors which are largely out of the control of the spraying machine and its operator such as wind speed and the surface characteristics of the spray target.

Spray chemicals are mixed and diluted with water for most applications, and different dilution rates and application rates are used for different spraying occasions. Both spray application rates and the sprayers are arbitrarily grouped as follows:

	Litres/ha
High volume	Over 600
Medium volume	200–600
Low volume	50–200
Very low volume	5–50
Ultra low volume	Less than 5

The same categories are used for bushes and trees but the application ranges are all higher.

Droplet formation

The spray liquid is broken down into the minute droplets required to give an adequate cover by:

1. forcing the liquid under pressure through a small orifice (hole) in a nozzle;
2. injecting the liquid into a fast-moving air stream;
3. spinning the liquid off the surface of a rapidly rotating disc or through a rapidly rotating circular gauze cage.

The first method is used by the majority of sprayers used in horticulture.

Types of sprayer

There are several types of sprayer in common used in horticulture.

1. *Knapsack and pedestrian-operated engine-driven sprayers* are usually of low volume and are used for a wide variety of purposes.
2. *Tractor-mounted or trailed ground-crop sprayers* may be of high, medium, or low volume according to their design and the purpose for which they are used.

3. *Orchard sprayers* are of high or low volume and are designed to work among trees and in soft-fruit plantations.

Knapsack sprayers

Knapsack sprayers are used for overall spraying of small areas, for spot treatment of small areas of weed, and for spraying individual shrubs and trees with fungicides and insecticides (Fig. 16.1).

Fig. 16.1 A hand-operated knapsack sprayer in action.
By permission of Cooper, Pegler & Co. Ltd.

These can be divided into three groups:
1. hand-operated pump sprayers;
2. pressurized, or pneumatic, sprayers;
3. engine-driven knapsack sprayers (See p. 141).

Hand-operated pump sprayers

These usually have 12 to 25 litre capacity spray containers made from brass, plastic, or stainless steel. Spray tanks made of plastic or stainless steel have the advantages that they are resistant to corrosion and are light to carry even when full of spray.

A hand-operated pump is attached to the spray container and this is worked by the operator carrying it. The spray is passed through a compression cylinder which evens out the liquid flow on each stroke of the pump. A hose is connected to the compression cylinder which leads to a hand lance or a short boom, and this can be fitted with a pressure gauge so that an accurate check can be kept on the pressure. A trigger on the hand lance controls the flow of spray and one, two, or sometimes three nozzles can be fitted at the end of the lance (Fig. 16.2).

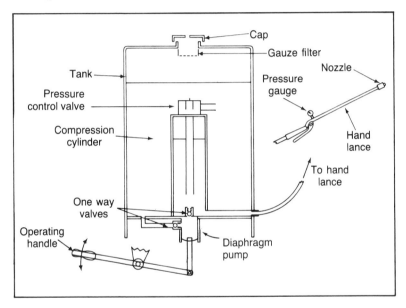

Fig. 16.2   A knapsack sprayer.

Spray is drawn from the tank by a downward movement of the diaphragm. The suction created opens valve A and closes valve B. As the diaphragm moves up on the return stroke it pressurizes the liquid which closes valve A and opens valve B. The spray then enters the compression cylinder and passes, still under pressure, to the hand lance. Maximum working pressure can be varied up to 5 bar by adjusting a pressure-release valve at the top of the compression cylinder. If the handle is pumped too slowly, or with short strokes, the pressure will be reduced (Fig. 16.3).

Some knapsack sprayers have the pump incorporated in the lance and are operated by backward and forward movements of a hand-piece which forms part of the lance. The pump provides pressure on both the forward and backward movements of the hand-piece so that a continuous spray is obtained.

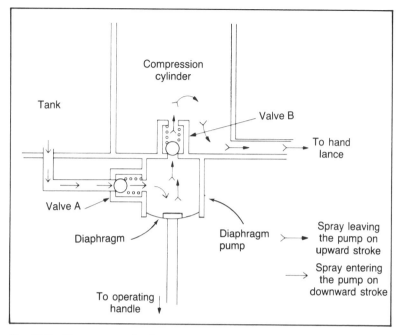

Fig. 16.3   The pump and valve arrangement of a knapsack sprayer.

Pressurized, or pneumatic, sprayers

These do not have to be hand-pumped during operation. There are some variations in design and method of operation but they can be divided into three groups.

In the first type, spray is put into the tank which is then pressurized by a built-in hand-pump. A safety valve prevents the tank from being over-pressurized. Pressing a trigger on the lance opens a valve to emit the spray. One pressurization may be enough to empty the tank.

In the second type, air is pumped into the tank until the minimum spraying pressure required is obtained. The spray is then pumped into the tank until the pressure recommended by the manufacturer is reached. The initial charge of air is retained by a spring-loaded valve when filling with spray so that it can be used to discharge completely all the spray in the tank.

The third type is similar to those in the previous group except that the spray is contained in a rubber or plastic bag inside the tank so that it does not cause corrosion. The spray is pumped into the bag until the total pressure is at the level recommended. The initial charge of air around the bag in the tank is retained almost indefinitely.

Pressurized sprayers are often mounted on trolleys and are fitted with short booms to give a 2 m or wider coverage when spraying larger areas such as lawns and golf greens.

Fig. 16.4   A ground-crop sprayer in action.
By permission of Lely Iseki Tractors.

Tractor-operated ground-crop sprayers (Fig. 16.4)

These sprayers have a large tank of 140 to 1 400 litres or more in capacity. Non-corrosive materials are widely used in the construction of these sprayers. A pump driven by the tractor power take-off draws the spray from the tank through a filter. On the outlet side of the pump there is a control valve which regulates the spray pressure, and hence the quantity of liquid sprayed, by bypassing a portion of the liquid pumped back to the tank. The liquid returned is used to agitate and mix the remaining spray in the tank. (Fig. 16.5). When operating at low pressure the valve is set so that most of the spray returns to the tank. At high pressure more of the spray is passed to the boom, but a proportion is still returned to the tank. The spray is delivered through a boom fitted with nozzles spaced at intervals of 225 to 450 mm. The length of the boom can vary to over 18 m, but shorter lengths are more common and are easier to fold up for transport.

A lever enables the spray to be shut off at the end of a run and redirects the spray from the pump to the tank. It may also cause 'suck back' of spray from the nozzles so that they do not dribble when the spray is shut off and cause a localized overdose of spray. Anti-drip nozzles may also be fitted to prevent nozzle dribble.

Pedestrian-operated ground-crop sprayers

These are smaller versions of the previous type of sprayer which are designed to apply herbicides to small areas such as larger lawns and golf greens. The sprayer is mounted on either one or two wheels and is pushed along by the operator. The pump is driven by a small engine.

Orchard sprayers

Most orchard spraying is carried out by 'mist blowers'. Very fine drop-

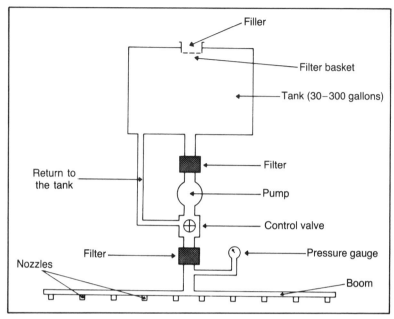

Fig. 16.5   The layout of a ground-crop sprayer.

lets are produced by spraying the liquid through nozzles or from a small, spinning, circular gauze cylinder into a high-speed air stream produced by a powerful fan. The mist of droplets produced is carried into the trees by the air blast which also stirs the foliage and ensures

Fig. 16.6   An orchard sprayer in action.
By permission of Drake and Fletcher Ltd.

that the trees are thoroughly covered in spray. The machines are either tractor-mounted or trailed, and both pump and fan are driven by the power take-off. (Fig. 16.6).

This spray can be directed in an arc to either side of the machine or to both sides simultaneously. Soft fruit can be treated as well as top fruit.

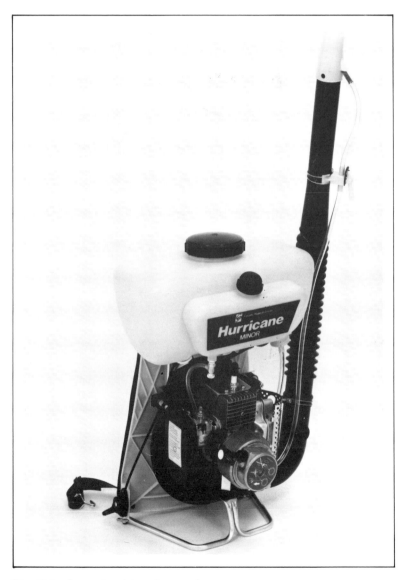

Fig. 16.7   An engine-driven knapsack sprayer.
By permission of Cooper, Pegler and Co. Ltd.

Engine-driven knapsack sprayers

These work on the same principle as the orchard sprayers described and are used for pest and disease control in smaller orchards and plantations. A small petrol engine incorporated in the sprayer drives the fan and the operator adjusts the outlet pipe to direct the spray as he walks along. This type of sprayer can also apply powders and granules (Fig. 16.7).

Nozzles

All nozzles are made of plastic, brass, stainless steel, or are fitted with ceramic tips.

The application rate for a given pressure, the size, and type of nozzle are indicated on the nozzle, or the nozzle may be colour coded to enable it to be easily identified.

The nozzles which are fitted to the spray boom or lance can vary in construction, but there are three types commonly in use.

Anvil nozzles

The construction and the working principles of these nozzles are shown in Fig. 16.8. They are used for the application of herbicides at low pressure and produce large droplets which reduce the chance of spray drift.

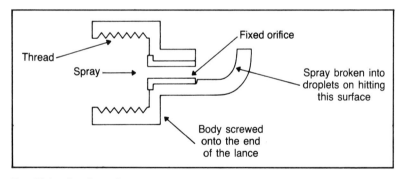

Fig. 16.8   Anvil nozzle.

Spray drift results when the spray is being broken up into particularly small drops and when the wind speed is high so that the drops are blown away from the area being sprayed. This may result in spray drifting on to a crop which is susceptible to the spray chemical. Thus spray drift must be minimized by spraying when there is little wind and by using the lowest spraying pressure which will give the correct spray pattern.

Fan nozzles

These are available in various sizes and are commonly used for applying herbicides. They consist of a body which is fastened to the boom or lance. Into the body fits a filter and a detachable nozzle tip. Both are held in the body by means of a securing nut. The nozzle

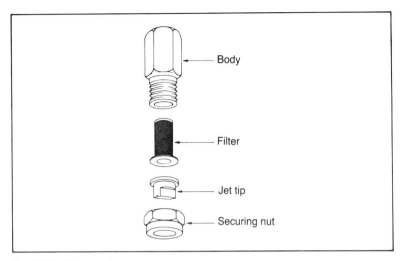

Fig. 16.9   Fan nozzle.

produces a flat fan of spray, and when they are fitted along a boom all the 'fans' must be in line. The height of the boom should be set so that the 'fans' cross over half-way between nozzle and target (Fig. 16.9).

Swirl nozzles

These are also available in various sizes and are commonly used to apply insecticides and fungicides.

Like the fan nozzles they consist of a body which is fastened to the boom or lance. The swirl plate and nozzle disc are separated by a washer to form a swirl chamber and are held in the body by a securing

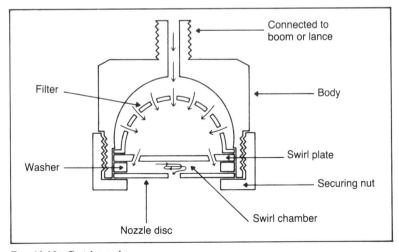

Fig. 16.10   Swirl nozzle.

143

nut. The swirl plate has a series of slanting holes drilled at a tangent through it so that as the spray passes through the holes it is given a swirling motion. The outer disc has a central hole in it, and the size of the hole together with the pressure govern the output from the nozzle. Considerable variations in contructional detail occur, but in every case the same features can be identified (Fig. 16.10).

The spray pattern produced by this type of nozzle is in the form of a hollow cone. Some swirl nozzles are designed to work at very low pressures so that drift hazards can be reduced.

Special booms

Special boom and nozzle arrangements to suit individual crops such as strawberries, cane fruit, vegetable and nursery crops can be obtained.

Dribble bars

This consists simply of a tube closed at both ends with a series of tiny holes drilled along its length. No pump is required as sufficient pressure is obtained by gravity by mounting the tank above the bar. Dribble bars can be fitted to the spout of a watering-can for spot application of sprays (Fig. 16.11).

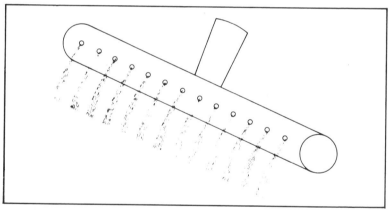

Fig. 16.11 A dribble bar.

Application rate

Application rates of sprayers should be easily adjusted so that any type of spray can be applied. The rate applied depends on the following:

1. *pressure.* The quantity of spray applied depends on the pressure; the higher the pressure the higher is the application rate, but higher pressures increase the proportion of smaller drops which are likely to drift. Higher pressure also widens the spread of drops from the nozzle. A lower pressure will minimize the risk of spray drift but reduces the application rate;
2. *nozzle size.* The bigger the orifice in the nozzle the greater the output. Tables are usually provided with the sprayer so that the

output from a particular nozzle at a given pressure can be found. If possible, it is better to fit a larger-sized nozzle to increase the application rate rather than to increase the pressure so as to minimize the danger of spray drift;

3. *forward speed*. In the case of knapsack sprayers this depends on the walking speed of the operator. Forward speed can be calculated by measuring how long it takes to walk a distance of 20 m.

Time taken to walk 20 m (s)	km/h
15	4.8
20	3.6
25	2.9
30	2.4

In the case of a tractor-operated sprayer the proofmeter is not always accurate enough to rely on when spraying, which is usually done at a speed of 6 to 8 km/h. Therefore the forward speed has to be checked by timing the tractor over a measured distance, usually 100 m.

Time taken to travel 100 m (s)	km/h
39	9
45	8
51	7
60	6

4. *Spraying width*. Width of spray of an individual nozzle is controlled by the operating pressure, its height above the ground, and the type of nozzle. If the pressure is constant, variation in height will alter the spraying width. A boom is usually at a fixed height above the ground to ensure that the spray pattern from adjoining nozzles match correctly to give an overall even distribution rate. The overall spraying width is the number of nozzles times the distance between any pair.

The spraying width of a knapsack sprayer can be measured by setting the sprayer to operate at the specified pressure and by observing the pattern on dry concrete or tarmac when the nozzles are held at the correct height.

Calibration of a knapsack sprayer

It is important that an operator should know what application rate he is applying with a sprayer at any setting. In order to obtain this rate the operator must be able to calibrate the sprayer. One method of calibration is as follows:

1. fit the correct type and size of nozzle as recommended by the manufacturer;
2. put some water into the sprayer and measure the effective spraying width in metres as described above;
3. mark out a distance of 20 m with canes, and time how long it takes to cover this distance at a steady walking pace;
4. multiply the distance travelled (20 m) by the effective spraying width to give the test area;

5. measure the amount of water sprayed from one nozzle in the same time that it took to walk the measured distance;
6. multiply the amount sprayed from one nozzle by the number of nozzles on the lance or boom.
7. the answer obtained will be the quantity of spray applied in litres to the test area. Dividing this volume by the area gives the application rate in litres per square metre;
8. measure the capacity of the tank in litres by filling it with water;
9. calculate the area that one tankful of spray will cover by dividing the capacity of the tank by the amount collected during the test and multiplying by the test area;
10. find out the rate at which the spray has to be applied from the spray chemical container or instruction leaflet;
11. multiply the area covered with one tankful of spray by the application rate per unit area of spray chemical to obtain the amount of spray chemical to be added to one tankful of water.

Example

1 and 2. The effective spraying width of a sprayer is found to be 1 m when held 0.6 m above the ground.
3. The time to walk the test distance of 20 m at a steady pace is 20 s.
4. Thus the test area covered is:

$$1 \text{ m} \times 20 \text{ m} = 20 \text{ m}^2$$

5. One nozzle applies 0.35 litre of spray in 20 s and there are two nozzles on the lance.
6. Thus the amount of spray applied in 20 s from the lance is:

$$0.35 \text{ litre} \times 2 \text{ nozzles} = 0.7 \text{ litre}$$

7. Therefore the rate of application is: 0.7 litre per 20 m^2.
8. The capacity of the sprayer is 14 litres.

9. Thus the area covered by one tankful of spray is: $\dfrac{14}{0.7} \times 20 = 400$ m^2.

10. The rate at which the chemical has to be applied is found to be 40 ml per 100 m^2.
11. Thus the amount of spray chemical to be added to a tankful of water is:

$$400 \times \frac{40}{100} = 160 \text{ ml}$$

If the application rate is given in litres or kilograms per hectare the procedure is the same; 1 ha = 10 000 m², and so the tank capacity is divided by 10 000.

Example

Application rate is 3 kg/ha. The area covered by one tankful is 400 m². Thus the amount of spray chemical to be added to a tankful of water

is: $400 \times \dfrac{3}{10\ 000} = 0.12$ kg, or 120 g

Calibration of a tractor-operated sprayer

Tractor-operated ground-crop sprayers can be calibrated in a similar manner except that the tractor is timed over a longer distance, perhaps 100 m, and the amount of water delivered during the test time is found by measuring water back into the tank to bring the level back to the starting level, which was marked, rather than by catching the liquid delivered by an individual nozzle. The engine speed at which standard power take-off speed is given must be used.

Example

1 and 2. The boom is fitted with 16 nozzles at 250 mm spacing.
  Thus the effective spraying width is: 16×250 mm $= 4$ m
3. The time over a measured distance of 100 m is 60 s.
4. The test area is 4 m $\times$ 100 m $= 400$ m².
5. After spraying for 60 s, 9 litres had to be replaced in the sprayer tank.

7. Therefore the application rate is: 9 litres per 400 m² or $\dfrac{9}{400} \times 10\,000$

$= 225$ litres/ ha

If this is not the required application rate the operating pressure must be adjusted slightly or a slightly different forward speed used and the calibration repeated until the desired application rate is achieved.

Spraying

The spray tank is always partially filled with water before the calculated amount of chemical is added. After mixing thoroughly the tank is filled up with water.

It is not advisable to spray if it is windy, i.e. a moderate breeze when small branches are moving, as this can cause uneven application and there is risk of spray drift. If rain in likely spraying can be wasteful and ineffective, as the spray will be washed off the plant on to the soil. Spraying on a hot day must also be avoided as some sprays can cause leaf scorch if they dry too quickly on the leaves, and some of the spray may be evaporated.

When spraying an area of ground, markers must be used to prevent overlap or missed areas. For knapsack spraying, canes or garden lines should be used to mark the spray bout widths and to help the operator walk in a straight line.

If the sprayer runs out in the middle of a run, mark the spot so that spraying can be restarted from that point so that no area is missed. One or two spare nozzles should be carried in a tin so that if a nozzle becomes blocked it can easily be exchanged. When the tank is refilled the blocked nozzle can be cleaned in water and be put back in the box for use if other nozzles become blocked. Never poke a nozzle with a piece of wire to clean it as this can damage it and spoil the spray pattern. On no account should a nozzle be inserted into the mouth in an attempt to clear a blockage in the nozzle by blowing through it.

A guard to prevent spray touching nearby trees or plants can be attached to some lances when using non-selective weedkillers.

After spraying

At the end of the day the tank must be drained of any spray and thoroughly washed out. If a knapsack sprayer is used, at least 10 litres of water should be pumped through the sprayer, and the addition of a detergent will help to remove all traces of the spray chemical. With a larger sprayer a greater quantity of water will be required. The water should be pumped out on to some waste land so that there is no risk of human or animal contamination.

If the sprayer is to be used the next day it can be left full of water overnight. The outside of the sprayer should also be washed down. The sprayer should also be thoroughly cleaned out when changing from one spray chemical to another.

At the end of a season the sprayer should be given an extra thorough wash-out with a solution of 0.5 kg of soda in 50 litres of water, followed by two or three rinsings of water.

The outside of the sprayer should be cleaned down and any parts where the paint has been worn off and corrosion could start should be cleaned and repainted.

The nozzles and the filters should be removed and cleaned. The pump and the hoses should be drained, after which the hoses should be removed for separate storage. The control valves should be slackened off and checked for wear. The hand lances should be removed and their condition checked.

All moving parts should be oiled and greased.

Spare parts should be ordered, or fitted if in stock, so that the sprayer is ready for use when it is next needed.

Safety

Most of the spray chemicals used are poisonous and can be harmful to the operator in very small doses. Extreme care is necessary when handling spray chemicals. Poisoning can take place by inhaling, skin contact or through the mouth. The following safety recommendations should be observed.

1. Make sure that you are familiar with the sprayer before using it.
2. Read the label carefully on the spray container and mix the spray in accordance with the maker's instructions.
3. Keep all spray chemicals under lock and key in a store designed for the purpose when not in use.
4. Rinse out and discard empty spray chemical containers by burying or burning them. Do not leave them lying about as they might get into the hands of children.
5. Wear rubber gloves at all times when handling spray chemicals and use the correct protective clothing if spraying any of the scheduled chemicals.
6. Always wash your hands and face thoroughly before eating.
7. If you feel ill when spraying, stop at once and report to a doctor telling him the type of spray being used.

8. Warn beekeepers nearby the day before you intend to spray.
9. Do not spray if it is windy.
10. Stop the machine before making any adjustments to it.
11. Avoid inhaling the spray.
12. Treat all chemicals as poisons.
13. Put up warning notices if spraying in a public place.

Questions

1. What is meant by low-, medium- and high-volume spraying?
2. Explain how one type of knapsack sprayer works.
3. Make a drawing to show the layout of components of a ground-crop sprayer.
4. Describe two types of nozzles used on sprayers.
5. How can spray drift be prevented?
6. List the factors affecting the application rates of sprayers.
7. A knapsack sprayer contains 18 litres of spray. The sprayer will cover an area 100 × 5 m with one tankful of spray when used by the operator. Calculate: (a) the number of tanks of spray required per hectare; (b) the quantity of spray chemical per tank if the rate of spray to be applied is 3 litres/ha.
8. Explain the procedures for cleaning a sprayer after use.
9. List the important safety factors which should be observed when using sprayers.

17 Lawn care equipment

As a result of the increase in grass areas devoted to recreational purposes, and the increase in labour costs, there has been an increase in the use of machinery in the practice of turf management. In this chapter a few of the mechanical aids now available to the groundsman are described.

Turf aerators

The purpose of turf aeration is to allow air to pass to the grass roots. It improves the surface drainage and enables top-dressings to be easily absorbed. With continual consolidation, especially during wet weather, this can be quite a problem on football pitches. Turf aerators can range from small hand-operated tools to larger tractor-mounted or trailed machines.

The hand aerator consists of a metal frame, with a wooden hand grip, on which four or five tines are fitted. The aerator is pushed into the ground with the foot and then pulled out so that the tines leave deep holes. It is only suitable for small areas such as small lawns or goalmouths.

For large areas self-propelled aerators can be used. (Fig. 17.1). These consist of a round cage on to which are fitted several rows of tines. The bars on to which the tines are mounted can pivot and are spring loaded so that they can enter and leave the turf cleanly and do not tear it. The cage is driven by a small petrol engine and carried on wheels for transport. To put the aerator into action the land wheels are raised by a lever. The cage of tines now rests on the ground providing the forward motion and directing the full weight of the machine onto the tines. On another machine pushing the machine forward transfers the weight from the land wheels and on to the cage. This gives good penetration and prevents wheel marks even if the ground is wet. Hydraulic rams can also be used to push the aerator tines into the turf. A box can be fitted behind to collect the cores of earth when using hollow tines. This removes the need to collect all the cores up afterwards and leaves the turf available for playing on immediately.

Larger tractor-operated aerators are in common use and can be either trailed or mounted. They consist of a series of four- or six-sided plates, equally spaced along a shaft. At each corner of every plate fixed or free-swinging tines are mounted and as the plates are pulled forward the tines enter the ground. If they are free swinging they enter and

Fig. 17.1　A turf aerator.

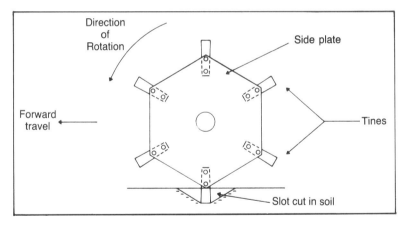

Fig. 17.2　The action of a tractor-drawn aerator.

leave the ground vertically, so reducing turf damage (Figs. 17.2 and 17.3).

These machines have a high rate of work. A self-propelled aerator 750 mm wide and containing 11 rows of 6 tines can produce approximately 320 000 holes per hour. An operator with a 5-tine hand aerator making a 100 holes per minute would take nearly 54 hours to make the same number of holes and cover the same area of turf. There are 4 common types of tine available (Fig. 17.4).

Fig. 17.3 A tractor-drawn aerator in action.
By permission of Sisis Equipment (Macclesfield) Ltd.

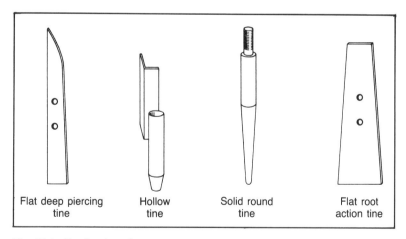

Flat deep piercing tine   Hollow tine   Solid round tine   Flat root action tine

Fig. 17.4 A selection of aerator tines.

1. *The flat deep-piercing tine.* This tine helps in soil aeration and cultivation.
2. *The hollow tine.* This takes out a core of soil and is used for turf renovation. The core of soil removed can be replaced by fresh soil brushed into the holes. The tines are self-cleaning as they are open at each end and the cores pass right through them.

3. *The solid round tine.* This tine spikes the ground and helps in the removal of surface water. The soil around the hole is compressed.
4. *The flat root-action tine.* This tine is used to prune the grass roots and encourage healthier growth.

Another type of aerator is available for aerating the soil 125 to 175 mm below the turf. Its action is rather like a mini subsoiler. The machine is self-propelled and has three vertical blades which cut into the turf. Mounted horizontally on the bottom of each blade there is a round piece of metal which acts as an expander. The blades oscillate as they are pulled through the soil and this has the effect of the expander cultivating the soil below the turf. This action helps to reduce compaction, improves drainage, and allows oxygen to reach the grass roots. A spring-loaded rear roller closes up the slot made by the blades, so leaving the turf undamaged and ready to play on.

Maintenance

The maintenance of engine-driven models should be carried out as described in Chapters 6, 7, and 8.

All the bearings and moving parts should be greased and a check made to see that there are no loose tines. Any that are loose should be tightened. Broken or bent tines should be replaced.

Rakes

Motorized rakes are designed to remove dead grass from turf areas. (Fig. 17.7). They also reduce the spread of trailing weeds and help to aerate the surface. A series of engine-driven, vertically rotating knives or cutters mounted along a shaft with spacers between them, rake through the grass, tearing out the dead grass (Figs. 17.5 and 17.6). They are designed to rotate at high speed and are adjustable in working depth. Machines may be either hand-pushed or self-propelled according to their size. Working widths can vary from 350 to 500 mm.

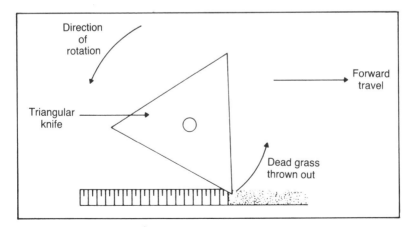

Fig. 17.5   The motor rake knife action.

Fig. 17.6   Vertical cutting knives of a motor rake.

The depth of operation can be controlled by adjusting rollers or wheels fitted at the front or rear of the machines. Lifting the rollers or wheels lowers the knives, and vice versa. The dead grass removed is thrown out in front of the machine and can be either left on the ground or collected in a box attachment fitted to the machine. The amount can be quite considerable on a turf area that has not been raked before. On the larger machines a rear roller is provided, while the hand-operated machines have two side wheels. Scrapers are usually provided for both the roller and the wheels. The knives are made out of hardened steel and can be easily replaced when they become worn or broken. Roller chains or vee belts provide the drive to the knives and roller.

Maintenance

Maintenance of engine-driven models should be carried out as described in Chapters 6, 7, and 8. All moving parts should be greased and belts or chain tensions should be checked and adjusted to the maker's specification.

Scarifiers

Scarifiers are used to provide a more gentle raking of the grass than that obtained with a motor-driven rake. Scarifying helps the entry of moisture, fertilizers and dressings into the turf. It also prevents the formation of moss by allowing air to circulate freely at the base of the grass shoots. Scarifiers have a series of galvanized wire tines which are mounted on a bar which is drawn across the grass or mounted on bars in a cross formation which are rotated by a small engine. The tines are hard wearing and their life is quite long. The width of the scarifier can

Fig. 17.7   A motor rake in action.
By permission of Sisis Equipment (Macclesfield) Ltd.

vary from a small 300 mm wide hand machine to a 1 500 mm attachment for use behind a tractor. Other types have the tines mounted on to a shaft which rotates horizontally and the tines come in contact with the turf as the shaft rotates.

Maintenance

Maintenance of engine-driven models should be carried out as described in Chapters 6, 7, and 8. The tines should be checked to ensure that they are mounted rigidly and have not worked loose.

Rollers

Rollers are frequently used for consolidation on cricket wickets, bowling-greens and golf-greens. They are also used to press down small stones on areas of newly seeded grass before mowing.

Rollers used for this purpose are of the flat type and are made of either steel or cast iron. They are available in a wide variety of widths and diameters. A good roller should consist of at least two sections, so that each section can rotate at different speeds when turning corners, to prevent damage to the grass. Rollers can be either hand-pushed, self-propelled, or tractor-drawn. Most rollers are provided with some form of ballasting so that extra weight in the form of concrete blocks, bags of sand, or cast iron can be added. Other rollers are provided with closed ends so that the ballast in the form of water, sand, or concrete can be put inside the roller. Water provides a good form of ballast, as it can be quickly removed and the roller weight can be changed very

155

easily. Very heavy rollers are used on cricket wickets, lighter ones on bowling-greens and golf-greens.

For large areas rollers can be pulled behind a tractor in gangs. Scrapers should be provided to prevent the build-up of soil on the roller, especially if the ground is damp.

Maintenance

The maintenance of rollers is restricted to the greasing of the roll bearings. These are often of the cast-iron type and are easily replaceable. Water-ballasted rollers must be drained before winter.

Brushes

Brushes are used in turf maintenance for brushing the grass to encourage aeration, scatter worm casts, and work in top-dressings. The bristles are made from whalebone, plastic, or hair. The working width of the brush can vary from 300 to 1 800 mm. They can be attached to small hand-pushed machines which are also adaptable for scarifying, piercing, or aeration. Larger brushes are attached to the rear of tractor-drawn machines.

Edge trimmers

With the large increase of ornamental lawns, greater lengths of lawn edges have to be trimmed. Powered edge trimmers have greatly reduced the time necessary for this from that taken by hand edging shears. The power supply for these machines is either a small petrol engine or a battery-operated motor. Edge trimmers are not usually self-

Fig. 17.8   An edge trimmer.

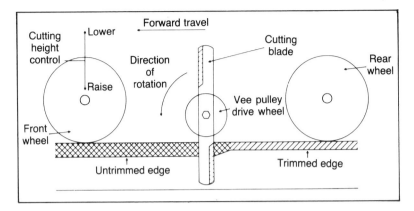

Fig. 17.9 The layout of an edge trimmer.

propelled and have to be hand-pushed, the power being supplied to
the blade only (Fig. 17.8).

The trimmer has a vertical blade which may be either mounted on
the shaft of the motor or driven by a small vee belt. The blade over-
hangs the edge of the lawn and rotates at high speed. The blade has
a sharp edge and cuts the grass in contact in the same way as a rotary
mower (Fig. 17.9).

Good lawn edges are essential if a neat job is to be made. The blade
on some machines can also be used on the horizontal, so that the
trimmer can be used as a small rotary mower for trimming in tight
corners. The battery-operated edger will trim about 1.6 km (1 mile) of
lawn edge per change, but this can be increased if larger batteries are
fitted. Included with the battery edgers, but not necessarily built in, is
a mains charger, which will ensure that the battery is fully charged,
ready for use.

Depth of cut is controlled by the front roller or wheel and by raising
the roller the depth of cut is increased. Care must be taken when using
these machines, that the blade does not come into contact with stones,
which will be flung up and may cause an accident. They also blunt the
blade edge. The blade is usually guarded.

Maintenance

Maintenance of engines should be carried out as described in Chapters
6, 7, and 8. All moving parts should be greased daily and the blade
checked for sharpness. Battery-operated edges must have the level of
electrolyte checked in each cell before charging and, if necessary, they
must be topped up with distilled water.

Turf cutters

Turf can be cut for relaying by hand, but turf cutters mechanize the
process and make the work very much easier. Turf cut by machine is
even in thickness and is much easier to lay.

Fig. 17.10   A turf cutter.
By permission of Allen Power Equipment Ltd.

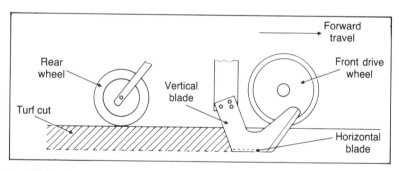

Fig. 17.11   Action of a turf cutter.

A small turf cutter consists of two L-shaped blades which are drawn under the turf at a depth which is adjustable. The depth of cut can usually be adjusted up to 75 mm deep and also the pitch of the blade can be altered to suit the depth. The turf width is determined by the distance between the vertical parts of the blades and is about 300 mm on a small machine but can vary up to 600 mm on larger machines (Figs. 17.10 and 17.11).

Turf cutters are driven by a single-cylinder petrol engine and the drive to the rubber-tyred roller is by vee belt and chain. The rear of the machine is supported by one or two rubber-tyred wheels. Some machines have an additional vertical blade which acts like a guillotine, working at right angles to the direction of travel and cutting the turf up into convenient lengths. After cutting the turf can be lifted and rolled. There are some machines designed to follow the turf cutter which lift, roll, and deposit the rolls of turf.

A small machine cutting a 300 mm turf can cut up to 10 m² of turf per minute.

Maintenance

Maintenance of engine-driven models should be carried out as described in Chapters 6, 7, and 8. The tension of the drive belts and chains should be adjusted as specified by the maker. All moving parts should be oiled or greased and the blades kept sharp or replaced if necessary.

Questions

1. Why are turf aerators used on lawn areas?
2. What types of aerator tine are available and what effect do they have on the turf?
3. Why are scarifiers used on lawn areas?
4. Describe the cutting mechanism of a lawn edger.
5. Make a drawing to show how a power-driven turf cutter works.

18 Chain saws and hedge cutters

Chain saws (Fig. 18.1)

Chain saws are used in horticulture to trim dead or diseased wood from ornamental trees, to remove inconveniently placed branches, and to fell trees.

The chain saw or power saw, as it is sometimes called, is a light, portable saw normally designed to be used by one person. Cutting is done by an endless chain fitted with cutters which run around a flat plate called the guide bar. The drive links of the chain run in a groove machined in the edge of the guide bar and they are pulled along by the teeth of a sprocket which engage them.

The sprocket in turn is driven at full engine speed by a small two-stroke petrol engine to which the guide bar is rigidly bolted, or, less commonly, by an electric motor or hydraulic power unit.

The engine

The engine must be of the two-stroke type so that the saw can be used

Fig. 18.1 A chain saw.

at any angle. This type of engine is lubricated by oil mixed with the petrol as described in Chapter 6, whereas the scoop or gear used in a four-stroke engine to flick oil on to the moving parts will not pick up sufficient oil if the engine is not level, resulting in oil starvation and possibly engine seizure. The engine also has a diaphragm-type carburettor to ensure a steady supply of petrol when the saw is being used on its side, or upside down when undercutting a branch. The engine is air cooled to help keep down weight and the air cleaner is usually a nylon gauze type which can easily be removed for cleaning. As the chain saw works in very dirty conditions the air passages around the engine cylinder and the air cleaner soon become blocked if not regularly maintained. Blocked air passages will result in the engine overheating, and a blocked air cleaner will lead to a shortage of air, causing lack of power.

Drive to the chain is transmitted through a centrifugal clutch mounted on the engine crankshaft which takes up the drive as the engine speed increases (Fig. 18.2). The throttle is usually of the trigger type and can easily be operated; increased pressure on it starts the chain running round the guide bar while releasing the trigger causes the chain to stop. A chain brake is often provided so that the operator can stop the drive to the chain quickly should the need arise. When force is applied to the operating lever a brake, in the form of a band, prevents the outer part of the centrifugal clutch from turning. On some chain saws the throttle trigger has two positions so that as well as allowing the engine to idle, the engine can also be stopped quickly and easily. Both the chain brake and the throttle trigger are two safety devices which allow the operator to stop the chain, the engine, or both in the case of an emergency.

The handles usually have rubber mountings to reduce vibration transmitted to the operator's hands. Excessive vibration can give rise to a complaint called 'white finger' in which the blood supply to the fingers is reduced by contraction of the capillaries causing the fingers to appear white. The handle should also be fitted with a guard to prevent the operator's hands becoming trapped by falling wood. The

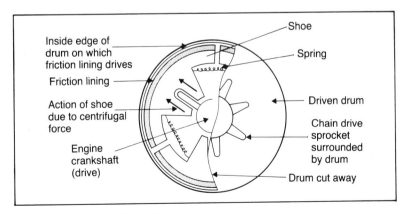

Fig. 18.2 Centrifugal clutch.

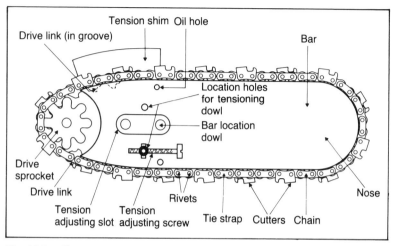

Fig. 18.3 Chain-saw guide bar and chain.

engines run at high speed, and although silencers are fitted chain saws tend to be rather noisy and it is advisable for the operator to wear earmuffs or plugs to prevent damage to the ears.

The chain

The chain is of the roller type and has left-hand and right-hand cutters spaced alternately along its length. In front of the cutters is a small projection called a depth gauge whose purpose is to control the depth of cut made by the cutter (Figs. 18.3 and 18.4).

Different types of chain are available. Chipper chains are general purpose and easy to sharpen and maintain. Micro-bit chains give a superior cutting action with quick chip removal and are more suitable

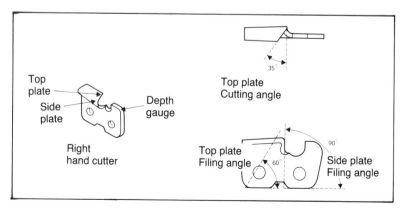

Fig. 18.4 Cutter parts and filing angles (chipper chain).

for cutting softwoods. Other chains are available for when hardwoods are to be cut.

Keeping the chain sharp and correctly tensioned is most important. Sharpening is usually done by hand. Special round files can be obtained for this purpose and are mounted in a special sharpening guide which can be clamped to the chain-saw guide bar. This ensures that as the file is pushed to and fro in the guide it is in the correct position to sharpen the cutters at the correct angle, to maintain the same angle on each cutter, and to remove the same amount from each cutter to obtain the sharpest edge. The cutters must always be filed from the inside outwards as the outer surfaces are chromed to harden them, and filing in the opposite direction will rapidly blunt the file.

Figure 18.4 shows typical sharpening angles on a general-purpose type chain. The height of the depth gauges must always be checked when sharpening and they must be filed down as the cutters wear, otherwise the depth of cut of the chain will be reduced causing the operator to press down harder on the saw, damaging the chain and guide bar.

Correct tension of the chain is also very important. A slack chain is likely to whip off the guide bar and could be very dangerous, and an overtight chain will wear rapidly. The chain is correctly tensioned when an upward pull on the chain in the centre of the guide bar fails to pull a drive link on the chain clear of the groove in the guide bar and the chain can be pulled round the guide bar by hand. Tensioning is achieved by moving the guide bar away from or towards the drive sprocket. This is done by slackening either one or two securing nuts, depending on the type of saw, turning the adjusting screw until the correct tension is obtained, and tightening the securing nuts, while all the time applying an upward pressure to the guide bar (Fig. 18.3).

The guide bar

The size of a chain saw is usually denoted by the length of the guide bar and common sizes range from 300 to 800 mm. The guide bar is made from steel, and as the greatest wear tends to take place at the end or nose of the bar the guide rails forming the sides of the groove there are hard-faced to slow down the rate of wear.

To reduce friction, roller or sprocket noses are fitted to carry the chain. These are often detachable so that they can be exchanged when worn. Wear is also greater on the lower side of the guide bar than on the top, and so each time the guide bar is removed for maintenance it is replaced the other way up to even out this wear.

The guide bar groove should be cleaned out every time the chain is removed and burrs caused by wear can be removed with a file, but ultimately the edges of the groove will become worn and sharp-edged and the guide bar will have to be replaced.

Lubrication of the chain is very important and so small holes are provided in the guide bar to allow oil to pass to the groove. Chain saws can either have manual chain lubrication, relying on the operator to pump oil to the chain during operation, or automatic lubrication in which oil is pumped by a small pump direct to the guide bar. A reservoir of oil is provided in the saw and this must be refilled each time the saw is refuelled. The groove and the oil holes must be cleaned frequently as the oil causes sawdust to clog and block the holes. The recommended chain-saw oil which has the correct viscosity must be used to ensure efficient lubrication of the chain in cold conditions. Failure to provide adequate lubrication will cause overheating, resulting in rapid wear of both chain and guide bar.

Using the saw

The chain saw is a very dangerous tool if not used correctly. Great care must be taken at all times by the operator, and spectators must be kept well away. For safe operation the following points should always be observed:

1. Always keep the cutters sharp.
2. Keep the chain correctly tensioned and never adjust it when hot or serious damage will be caused when the chain contracts on cooling.
3. Never walk along holding the saw with the engine running, and stop it whenever it is not to be used again immediately.
4. Never use the feet to hold timber when sawing.
5. Wear steel-capped boots when using a chain saw to protect the feet.
6. When starting the saw place it on the ground, put a foot through the saw handle and hold the handle firmly with one hand. Operate the recoil starter with the other hand.
7. Check that the chain saw will idle without driving the chain.
8. Use the felling spike provided in front of the handle to lever the saw through thick timber.
9. Always wear ear-plugs and a safety helmet with visor when sawing.

Maintenance

To keep the saw in correct working order the following maintenance must be carried out.

1. The engine must be regularly maintained as described in Chapters 6, 7, and 8. It is particularly important to keep the air cleaner clean and keep the fins on the cylinder head free of sawdust.
2. The chain must be kept sharp and correctly tensioned.
3. The groove and oilways must be kept clear of sawdust.
4. The depth guides on the chain must be checked and correctly set.
5. The nose-wheel bearing must be greased if fitted.
6. The chain oil lubrication reservoir must be kept topped up.

Hedge cutters

There are many types of hedge cutter on the market. They can be driven by small petrol engines, compressed air, or are electrically powered, the latter being by far the more common. The electrical supply can be from a 12 V battery – operated from the battery on a tractor or a small 12 V d.c. portable generator – or a 240 V mains supply if one is conveniently situated. A small domestic type with a rechargeable battery is also available. Cutting takes place between two blades, one of which reciprocates in close contact with a stationary one at a rate of between 33 and 66 strokes per minute. (On some models both blades reciprocate to provide the cutting action.) It will be noted that there are no razor-sharp edges on the cutting parts, but good clean undamaged edges with angles of just less than 90 degrees are required if good cutting is to be achieved (Fig. 18.5).

Some hedge cutters have an extension on the end of the moving blade like a small saw which can be used to saw through branches that are too thick to be cut by the blades. Hard metallic objects caught in the blade will damage the cutting edge and cutting will be impaired.

Fig. 18.5   Portable hedge cutter in action.
By permission of Allen Power Equipment Ltd.

Maintenance

Maintenance consists of keeping the edges in good condition and lubricating moving parts.

Sharpening the blades is a skilled job and is best left to an expert who has the equipment to ensure that the correct angle is maintained. A few drops of oil should be applied to the blades and the reciprocating mechanism and, if the latter is contained in a gearbox, topping up with the correct grade of oil will be required.

Care must be taken to prevent oil getting on to the brushes in the electric motor. Maintenance of the petrol engine on engine-driven models should be carried out as described in Chapters 6, 7, and 8. If mains electrical hedge cutters are used the earth lead must be checked frequently to ensure that any stray currents can be conducted safely to earth should a fault occur in the cutter. If it is wired from a 13 A three-pin plug the size of fuse recommended by the manufacturer must be fitted in the plug so that it will blow if the cutter is overloaded.

Questions

1. What are the two important checks to be made on the chain of a chain saw?
2. Why are two-stroke engines always used on chain saws?
3. What checks are made to indicate that the chain is correctly tensioned?
4. List the safety precautions that the operator of a chain saw must take when using a saw.
5. What is the purpose of the depth guide on a chain-saw cutter?
6. Why must the chain lubrication system be checked regularly?
7. What maintenance does a hedge cutter require?
8. What safety checks should be made on a mains electric hedge cutter?

19 Glasshouse equipment

In recent years there has been an increase in the amount of equipment installed in glasshouses to enable the artificial environment provided to be used to its fullest advantage. In this chapter a number of the common items of equipment are described.

Mist propagation (Fig. 19.1)

The technique of mist propagation enables cuttings of plants to be rooted easily. If the cuttings are kept in a humid atmosphere with a supply of warmth at the place where the roots will form, the process of root formation is greatly accelerated. The mist propagation unit supplies a fine mist of water droplets on to the leaves of the cuttings at frequent intervals. Below the cuttings bottom heat may be supplied by a soil-warming cable.

For a mist unit to work satisfactorily the glasshouse must have a clean

Fig. 19.1   A mist-propagation unit.

water supply at an adequate constant pressure and a properly installed mains electricity supply.

If the water pressure is too low a header tank placed in the roof of the glasshouse and a pump to supply it may have to be installed. This also provides a break between the mains supply and the mist unit which may be required by local water authorities.

Water entering the unit first passes through a filter in the Y-shaped section of pipe. This will remove any particles of dirt from the water and will prevent the small orifices in the nozzles producing the mist from becoming blocked. The filtered water passes on to the solenoid valve which controls the flow to the nozzles. The solenoid valve is electrically operated. It opens when a small current is passed to it from the control unit, and closes again when the current stops flowing. (Fig. 19.2).

When the solenoid valve is open, water passes on through the pipe to the nozzles. These are made of brass, plastic, or stainless steel. The jet of water leaving the nozzle hole strikes an anvil and breaks down into a mist of droplets fanning out all round the nozzle. The distance between hole and anvil is adjustable and controls the size of droplets in the spray mist. A range of nozzles is available to cover different areas. The nozzles may be mounted on standpipes projecting vertically approximately 450 mm above the rooting medium, or on supply pipes angled up from a main installed at the side of the bench. Alternatively, the nozzle may be set on a high-level horizontal pipe suspended from the roof trusses of the glasshouse.

The signal to operate the solenoid valve so that misting can begin is provided in many systems by an 'electronic leaf' placed among the cuttings. The top of the electronic leaf contains two electrodes mounted in insulating plastic which are connected to the control unit by a twin-core cable. When the top of the leaf is covered with a film of moisture a small current can flow between the electrodes, completing the circuit through the cable back to the control unit, and this is the signal which causes the solenoid valve to close. When the moisture on the leaf evaporates the circuit through the leaf is broken and the solenoid valve reopens. Once the leaf and, of course, its surroundings, become moist again the solenoid valve recloses, and misting ceases. By placing the leaf near to the nozzles the length of time the spray operates is reduced, and by placing it away from the nozzles towards the edge of the area being misted the length of time is increased.

Some modern control units are equipped with 'weaners'. The weaner control enables the time between the bursts of mist to be increased as the roots develop, reducing the humidity around the plants ready for when they are removed. This lessens the shock which occurs if cuttings are removed from the mist abruptly.

A solarimeter can be used to control a mist-propagation unit instead of the artificial leaf. Bright conditions accelerate transpiration losses which must be offset by more frequent bursts of mists. On dull days less moisture is needed. A solarimeter registers the variation in light energy falling on it and increases or decreases the number of mist periods in relation to this.

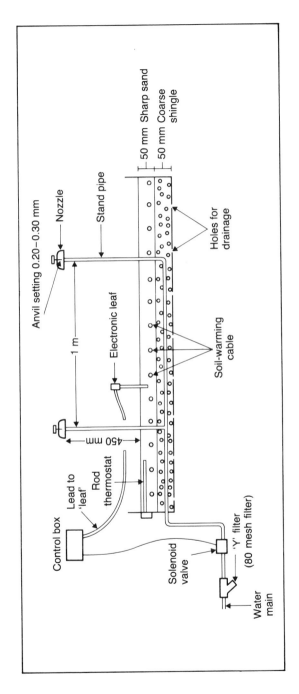

Fig. 19.2 A typical mist-propagation unit.

Maintenance

1. The top of the electronic leaf must be cleaned regularly to prevent the build-up of impurities which will cause a resistance to the flow of electricity. Dipping the leaf in diluted hydrochloric acid or rubbing it gently on fine glasspaper is recommended. On no account must the leaf be cleaned with the fingers, or the electrodes will be contaminated with a greasy layer.
2. The nozzles may require cleaning if deposits build up in them. Care must be taken not to damage the orifice of the nozzle, or the coverage of the spray mist may be affected.
3. The gauze in the Y filter will require cleaning from time to time.
4. All electrical wiring must be checked regularly for cracked insulation and poor connections.

Soil-warming cable

As the name implies, soil-warming cable is used for warming the soil, but it can also be used for warming the rooting medium in mist-propagation units, as described on p. 167, and also for air heating in some propagation frames. Soil-warming cables consist of plastic-covered resistance wires which warm up when an electrical current is passed through them, but the temperature reached is too low to melt the plastic. Two layouts are common. In one, a continuous circuit of single-core cable is laid out in loops starting at a control box and returning to it. In the other, the cable also contains within it the return wire necessary to complete the circuit. Both types of cable are of a predetermined length and must not be shortened, otherwise the loading will be changed and, in the second type, the return wire cut.

Soil-warming cables are available in various 'set' lengths and are usually operated from a 240 V mains supply, although some low-voltage cables operated from a transformer are available. The cable will have a specified loading given in watts, and by dividing the number of watts by the length of the cable the loading in watts per metre length can be found. The amount of heat, and hence the temperature required, are obtained by installing a cable supplying the number of watts per square metre recommended for a given application. Mist-propagation benches have a recommended loading of 162 W/m², and bench warming for propagation frames 80 W/m². Most soil-warming cable layouts incorporate a rod-type thermostat in the circuit which switches off the current to the cable when the required temperature has been reached, and switches it on again when the temperature has fallen. This allows a more accurate temperature to be obtained and reduces the running costs (Fig. 19.3).

Maintenance

Good watertight connections are essential with electrical equipment operating in the damp conditions found in glasshouses. As most cables are sealed units they cannot be repaired by the grower if failure occurs and must be returned to the maker or discarded. On no account must they be opened or shortened and reconnected.

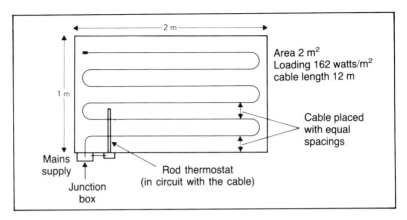

Fig. 19.3 Layout of soil-warming cable.

Dilutors

Dilutors are used to introduce concentrated solutions of soluble fertilizer, which are called stock solutions, into irrigation systems, particularly trickle systems, at predetermined dilution rates to form liquid feeds. Although different types are available, probably the commonest is the displacement dilutor, or charge dilutor as it is sometimes called.

The dilutor is connected into the pipeline so that the water going to the watering-points flows through it and, in so doing, passes through a form of venturi incorporated within it which is not unlike that in a carburettor. Two small tubes or holes bored in the dilutor, one at the venturi and one on the upstream side of it, connect with a container which is completely full of the concentrated stock solution of fertilizer. As the water flows through the venturi, a pressure difference develops between the two holes resulting in a proportion of the water passing down into the container through the connection on the upstream side, and displacing a corresponding volume of stock solution into the venturi through the other hole. This mixes with the main flow of water in the pipes on the way to the irrigation points and forms the liquid feed (Fig. 19.4).

It is important that the water entering the container is not allowed to mix with the stock solution. If it does the concentrated stock solution will be diluted before it is introduced into the main water flow at the venturi, the dilution rate will change, and the solution fed to the plants will become weaker as the level of stock solution in the container falls. This may not always matter, but must be avoided with plants in containers. The stock solution is denser than water, and so the water entering tends to float on top of it. To ensure that intermixing does not take place, the water entering is fed gently over a circular disc which distributes it evenly across the top of the liquid in the container, and the mouth of the pipe leading to the venturi through which the stock solution is forced is close to the bottom of the container. A fine gauze filter is fitted over the mouth of this pipe to prevent any particles entering. However, care

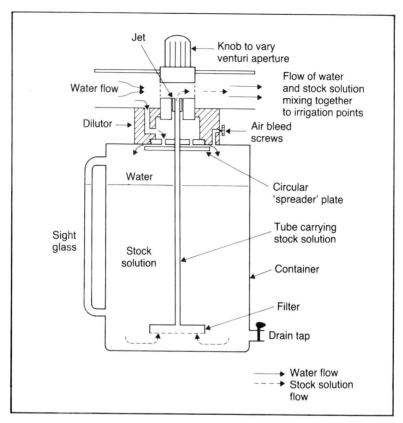

Fig. 19.4 A displacement dilutor.

must be taken to avoid shaking the container, or the water and stock solution will mix and the dilution rate will change. If this happens by accident the container will have to be emptied and refilled with fresh stock solution.

The containers, which may vary in size from 9 to 160 litres or more, have a sight glass fitted on the side, and the liquid feed is coloured so that it is easy to see when it is exhausted. When this happens the container must be emptied of water and be refilled completely with stock solution. An air-release screw is provided in the top of the container which can be opened to allow all the air out and to ensure that the container is completely full, after which it is closed again.

At the same time to avoid possible blockage the calibration jet should be taken out and thoroughly washed before replacing.

Calibration

The dilution rate is usually altered by changing a jet in the pipe leading to the venturi, and by varying the size of the venturi.

The dilutor is calibrated by collecting the water flowing through it in a vessel of known size – one of the smaller dilutor containers is ideal – timing how long it takes to fill, and then calculating the rate of water flow per hour. Alternatively, the water can be collected from one watering-point for a known time, and the total volume flowing through the dilutor calculated by multiplying this by the number of watering-points.

The manufacturer provides a calibration chart showing the size of jet, and the venturi setting necessary for different rates of water flow. Having set these the flow through the dilutor should be rechecked.

Many growers standardize on one dilution rate, and thus one dilutor setting for a particular installation, by modifying the stock solution analysis to suit.

Questions

1. What is the function of the solenoid valve in a mist-propagation unit?
2. Describe how a mist jet works, and explain how it can be adjusted to vary the mist produced.
3. What maintenance does the artificial leaf require?
4. Explain the difference between the two types of soil-warming cable available.
5. What would be the result of shortening a soil-warming cable?
6. What will be the result if water and stock solution in the container of a dilutor are allowed to mix?
7. What adjustments can be made to a dilutor to alter the dilution rate?
8. Describe the procedure for calibrating a dilutor supplying liquid feed through a trickle irrigation system to pot plants.

Metric equivalents

Although SI units have been used in the text the following equivalents may still prove to be useful.

Length	1 in	= 2.54 cm
	1 ft	= 30.48 cm (0.3048 m)
	1 yd	= 91.44 cm (0.9144 m)
	1 mile	= 1.6093 km
Area	1 acre	= 0.4047 hectares (ha)
	1 sq. yd	= 0.83613 m^2
Weight	1 oz	= 28.350 g
	1 lb	= 0.4536 kg
	1 cwt	= 50.8032 kg
	1 ton	= 1.0161 tonnes (t)
Capacity	1 gal	= 4.5461 litres (l)
Application rates	1 oz/sq. yd	= 33.903 g/m^2
	1 cwt/acre	= 125.54 kg/ha
	1 pint/acre	= 1.4042 l/ha
Pressure	1 lbf/sq. in	= 6.895 kN/m^2 = 0.06895 bar

Index